Gas Chromatography

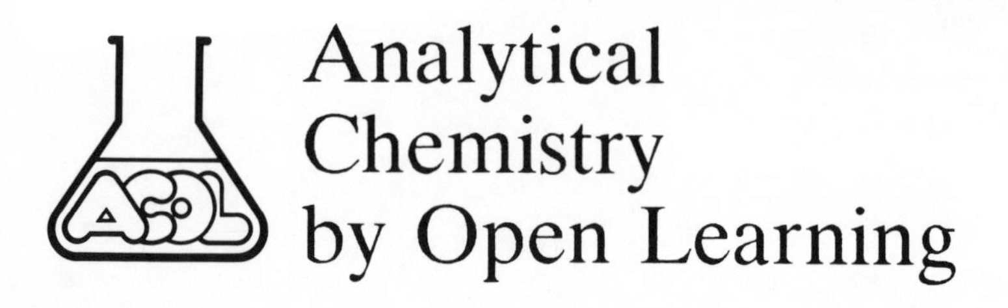

Project Director
BRIAN R CURRELL
Thames Polytechnic

Project Manager
JOHN W JAMES
Consultant

Project Advisors
ANTHONY D ASHMORE
Royal Society of Chemistry

DAVE W PARK
Consultant

Administrative Editor
NORMA CHADWICK
Thames Polytechnic

Editorial Board
NORMAN B CHAPMAN
Emeritus Professor, University of Hull

BRIAN R CURRELL
Thames Polytechnic

ARTHUR M JAMES
Emeritus Professor, University of London

DAVID KEALEY
Kingston Polytechnic

DAVID J MOWTHORPE
Sheffield City Polytechnic

ANTHONY C NORRIS
Portsmouth Polytechnic

F ELIZABETH PRICHARD
Royal Holloway and Bedford New College

Titles in Series:

Samples and Standards
Sample Pretreatment
Classical Methods
Measurement, Statistics and Computation
Using Literature
Instrumentation
Chromatographic Separations
Gas Chromatography
High Performance Liquid Chromatography
Electrophoresis
Thin Layer Chromatography
Visible and Ultraviolet Spectroscopy
Fluorescence and Phosphorescence Spectroscopy
Infra Red Spectroscopy
Atomic Absorption and Emission Spectroscopy
Nuclear Magnetic Resonance Spectroscopy
X-Ray Methods
Mass Spectrometry
Scanning Electron Microscopy and Microanalysis
Principles of Electroanalytical Methods
Potentiometry and Ion Selective Electrodes
Polarography and Other Voltammetric Methods
Radiochemical Methods
Clinical Specimens
Diagnostic Enzymology
Quantitative Bioassay
Assessment and Control of Biochemical Methods
Thermal Methods
Microprocessor Applications

Gas Chromatography

Analytical Chemistry by Open Learning

Author:
JOHN WILLETT
Wolverhampton Polytechnic, UK

Editor:
DAVID KEALEY

on behalf of ACOL

543.0896
W669g

Published on behalf of ACOL, London
by
JOHN WILEY & SONS
Chichester · New York · Brisbane · Toronto · Singapore

Barton College Library
Wilson, N.C. 27893

© Crown Copyright, 1987

Reprinted November 1988
Reprinted May 1989
Reprinted June 1990

Published by permission of the Controller of
Her Majesty's Stationery Office

All rights reserved.

No part of this book may be reproduced by any means, or transmitted, or translated into a machine language without the written permission of the publisher.

Library of Congress Cataloging in Publication Data:

Willett, John, 1938–
Gas chromatography.

(Analytical chemistry by open learning)
'Published on behalf of ACOL, London.'
1. Gas chromatography—Programmed instruction.
2. Chemistry, Analytic—Programmed instruction.
I. Kealey, D. (David) II. ACOL (Firm : London, England) III. Title. IV. Series.
QD79.C45W55 1987 543'.0896 86-28263
ISBN 0 471 91331 6
ISBN 0 471 91332 4 (pbk.)

British Library Cataloguing in Publication Data:

Willett, John
Gas Chromatography.—(Analytical chemistry)
1. Gas chromatography
I. Title II. Kealey, D. III. Analytical Chemistry by Open Learning/Project
IV. Series

543'.0896 QD79.C45

ISBN 0 471 91331 6
ISBN 0 471 91332 4 Pbk

Printed and bound in Great Britain by
Biddles Ltd, Guildford and King's Lynn

Analytical Chemistry

This series of texts is a result of an initiative by the Committee of Heads of Polytechnic Chemistry Departments in the United Kingdom. A project team based at Thames Polytechnic using funds available from the Manpower Services Commission 'Open Tech' Project have organised and managed the development of the material suitable for use by 'Distance Learners'. The contents of the various units have been identified, planned and written almost exclusively by groups of polytechnic staff, who are both expert in the subject area and are currently teaching in analytical chemistry.

The texts are for those interested in the basics of analytical chemistry and instrumental techniques who wish to study in a more flexible way than traditional institute attendance or to augment such attendance. A series of these units may be used by those undertaking courses leading to BTEC (levels IV and V), Royal Society of Chemistry (Certificates of Applied Chemistry) or other qualifications. The level is thus that of Senior Technician.

It is emphasised however that whilst the theoretical aspects of analytical chemistry can be studied in this way there is no substitute for the laboratory to learn the associated practical skills. In the U.K. there are nominated Polytechnics, Colleges and other Institutions who offer tutorial and practical support to achieve the practical objectives identified within each text. It is expected that many institutions worldwide will also provide such support.

The project will continue at Thames Polytechnic to support these 'Open Learning Texts', to continually refresh and update the material and to extend its coverage.

Further information about nominated support centres, the material or open learning techniques may be obtained from the project office at Thames Polytechnic, ACOL, Wellington St., Woolwich, London, SE18 6PF.

FEB 12 1991

How to Use an Open Learning Text

Open learning texts are designed as a convenient and flexible way of studying for people who, for a variety of reasons cannot use conventional education courses. You will learn from this text the principles of one subject in Analytical Chemistry, but only by putting this knowledge into practice, under professional supervision, will you gain a full understanding of the analytical techniques described.

To achieve the full benefit from an open learning text you need to plan your place and time of study.

- Find the most suitable place to study where you can work without disturbance.
- If you have a tutor supervising your study discuss with him, or her, the date by which you should have completed this text.
- Some people study perfectly well in irregular bursts, however most students find that setting aside a certain number of hours each day is the most satisfactory method. It is for you to decide which pattern of study suits you best.
- If you decide to study for several hours at once, take short breaks of five or ten minutes every half hour or so. You will find that this method maintains a higher overall level of concentration.

Before you begin a detailed reading of the text, familiarise yourself with the general layout of the material. Have a look at the course contents list at the front of the book and flip through the pages to get a general impression of the way the subject is dealt with. You will find that there is space on the pages to make comments alongside the text as you study—your own notes for highlighting points that you feel are particularly important. Indicate in the margin the points you would like to discuss further with a tutor or fellow student. When you come to revise, these personal study notes will be very useful.

Π When you find a paragraph in the text marked with a symbol such as is shown here, this is where you get involved. At this point you are directed to do things: draw graphs, answer questions, perform calculations, etc. Do make an attempt at these activities. If necessary cover the succeeding response with a piece of paper until you are ready to read on. This is an opportunity for you to learn by participating in the subject and although the text continues by discussing your response, there is no better way to learn than by working things out for yourself.

We have introduced self assessment questions (SAQ) at appropriate places in the text. These SAQs provide for you a way of finding out if you understand what you have just been studying. There is space on the page for your answer and for any comments you want to add after reading the author's response. You will find the author's response to each SAQ at the end of the text. Compare what you have written with the response provided and read the discussion and advice.

At intervals in the text you will find a Summary and List of Objectives. The Summary will emphasise the important points covered by the material you have just read and the Objectives will give you a checklist of tasks you should then be able to achieve.

You can revise the Unit, perhaps for a formal examination, by re-reading the Summary and the Objectives, and by working through some of the SAQs. This should quickly alert you to areas of the text that need further study.

At the end of the book you will find for reference lists of commonly used scientific symbols and values, units of measurement and also a periodic table.

Contents

Study Guide

This unit is intended to provide you with a working knowledge of gas chromatography. It will not turn you into a fully experienced gas chromatographer – only months and years of practice can do that, for gas chromatography is still something of an art. Many of us have learned through standing in front of a gas chromatograph and making endless mistakes. It is very frustrating, but it is probably the best way to learn to recognise the symptoms of the things which can go wrong with gas chromatography. There are many settings on a gas chromatograph which can be wrongly chosen, and many *physical* malfunctions that have nothing to do with the chemistry of the process but affect the outcome of an analysis dramatically. They have to be experienced, or at least described to you before you learn to recognise the symptoms and diagnose the faults. This unit will try to provide you with the equivalent of that experience by adopting the standpoint of a practising gas chromatographer.

Even so, whilst emphasising the practical side of the subject, the theoretical side will not be ignored. It is only by understanding the processes occurring in your instrument that you will get the best out of it. The purpose of considering the theory, though, is quite clearly to improve your practical performance.

I will assume that you have an understanding of chemistry equivalent to that of a student who has passed HNC or HTC in chemistry (BTEC), and a knowledge of physics going to at least GCE(OL).

We shall need to talk especially about chemical equilibria, the properties of solutions and of gases, the structures of molecules and how they attract one another, and we shall also have to use your knowledge of elementary electricity. It will be assumed that you have some basic knowledge of Chromatography, possibly by having studied the Unit of this ACOL series *Chromatographic Separations*.

You may find that another author's views on gas chromatography will clarify, for you, some aspects of this text. I hope that this will not be necessary too often, but if it is, I have always found Stock and Rice's *Chromatographic Methods* (now revised by Braithwaite and Smith) very useful and for more advanced work, Ettre and Zlatkis' *Gas Chromatography* takes a lot of beating. Pattinson has produced an interesting programmed learning text which you might find helpful.

Inevitably, in a distance learning package you will not get anything like enough *hands on* practical experience, but at your regional centre you may have the opportunity to use gas chromatographs and computer controlled simulations which will help you along the road to becoming an experienced gas chromatographer.

Supporting Practical Work

1. GENERAL CONSIDERATIONS

Gas chromatographs are quite widely available, so it is quite likely that you will be able to gain practical experience at your place of work. If you do this, you should take steps to make friends with your 'in-house' gas chromatographer, for he will be able to guide you through the experiments suggested below and will also be able to introduce you to the design idiosyncracies of the particular gas chromatograph which you will use. (There are many manufacturers of gc's, and although they must all have more or less the same facilities, each manufacturer has his own approach. Furthermore, most instruments are pretty robust, and there are many early instruments still in use and doing valuable work as well as very modern and complex examples. There is a case for recommending that, if you can, you try to use an early, basic instrument so that you have to do everything for yourself and so learn a little more about basic principles than you will do if you work on a fully automated gc).

2. AIMS

(*a*) To provide a basic experience of using a gas chromatograph and associated equipment.

(*b*) To illustrate the effect on separation by gas chromatography of the various parameters under the operator's control.

(*c*) To illustrate the feasibility of qualitative and quantitative analysis of volatile and non-volatile mixtures by gas chromatography.

3. SUGGESTED EXPERIMENTS

(*a*) The examination of the effect of stationary phase, column length, temperature, carrier gas flow rate, sample size etc upon the separation of benzene, cyclohexane and ethanol.

(*b*) The determination of the concentration of ethanol in a dilute aqueous sample in the region relevent to drink-driving legislation (approximately 100 mg/100 ml) using propan-1-ol as internal standard.

(*c*) The identification of an acrylate polymer by pyrolysis gas chromatography (a bit smelly, if you use the monomers to determine reference retention times, but worth doing).

(*d*) The estimation of the composition of inhaled and exhaled air, using a katharometer (TCD).

Bibliography

1. (a) F. W. Fifield and D. Kealey, *Principles and Practice of Analytical Chemistry*, 2nd Ed., International Text-book Co, 1983.

   (b) D. A. Skoog, *Principles of Instrumental Analysis*, 3rd Ed., Saunders College Publishing, 1985.

   (c) H. H. Willard, L. L. Merritt, J. A. Dean, F. A. Settle, *Instrumental Methods of Analysis*, 6th Ed., Wadsworth Publishing Co., 1981.

2. (a) R. Stock and C. B. F. Rice, *Chromatographic Methods*, 3rd Ed., Chapman and Hall, London.

   (b) A. Braithwaite and F. Smith, *Chromatographic Methods*, 4th Ed., Chapman and Hall, London, 1985.

3. L. S. Ettre and A. Zlatkis, *The Practice of Gas Chromatography*, Interscience, New York, 1967.

4. J. B. Pattinson, *A Progammed Introduction to Gas Liquid Chromatography*, 2nd Ed., Heyden and Son Ltd., London, 1969.

5. A. J. P. Martin and R. L. M. Synge, *Biochem, J.*, **35**, 1358 (1941).

6. A. T. James and A. J. P. Martin, *Biochem, J.*, **50**, 679 (1952).

7. M. J. E. Golay, *Gas Chromatography*, 1960 (Ed. R. P. W. Scott), Butterworths, Washington, 1960.

8. S. H. Lamger, and R. J. Sheehan, *Anal. Chem. and Inst.*, **6**, 298 (1968).

9. R. L. Levy, *Chromatographic Reviews*, **8**, 48 (1966).

10. R. W. May, *Chemical Society Analytical Sciences Monograph No 3*.

11. E. Kovats, *Helv. Chim. Acta.*, **41**, 1915 (1958).

12. A. Zlatkis and V. Pretorius, *Preparative Gas Chromatography*, John Wiley and Sons, Ltd., New York, 1971.

13. F. W. McLafferty, *Science*, **151**, 641 (1966).

14. R. Ryhage, *Anal. Chem.*, **36**, 759 (1964).

15. W. McFadden, *Techniques of Combined Gas Chromatography/Mass Spectrometry: Applications in Organic Analysis*, John Wiley and Sons Ltd., New York (1973).

NOTE References 1 a, b and c are books containing chapters on gas chromatography.

References 2, 3, 4 and 7 are more specialised texts.

1. Introduction

From the knowledge of gas chromatography which you have picked up informally and discovered by earlier studies, try to answer the question below.

Π Complete the sentences A and B by pairing with each of them a *second half* from the list (*i*) to (*vi*).

A. Gas chromatography differs from other chromatographic techniques because it ...

B. Gas chromatography would be a good way to analyse ...

(*i*) uses a detector

(*ii*) uses a gas as a mobile phase

(*iii*) uses liquid stationary phases

(*iv*) the contents of a camping gas cylinder

(*v*) a sample of iron ore

(*vi*) a bag of fertiliser

If you paired A with (*ii*) and B with (*iv*), well done! (*i*) could not be the right pairing for A, since high performance liquid chromatography also uses a detector, and neither could (*iii*) be right, since liquid stationary phases are also used on occasions in high performance liquid chromatography, thin-layer chromatography and paper chromatography. Nor could (*v*) or (*vi*) be the right pairing for B since gas chromatography needs to get the analyte into the gas phase for it to pass through the column and neither iron ore nor fertiliser are volatile. If you didn't get it right, don't worry. This Unit will soon help you to understand why.

Gas chromatography has a history going back to about 1940. That such a technique might be possible was suggested by Martin who did so much to establish chromatography as a valuable tool for the analyst. Several reports of rudimentary separations by GC appeared during the 1940's but it was not until the early 1950's that the problems had been overcome sufficiently for it to become a useful technique. This can almost be dated from the publication by Martin and James[5] of a description of gas–liquid chromatography, which has the reliability needed for routine analysis. The first commercial gas chromatograph (manufactured by Griffin and George) appeared in the mid-fifties, to be followed by the products of Pye, Perkin Elmer, Wilkins (Varian), Hewlett Packard and a host of others. The availability of these instruments led to the wide use and rapid development of gas chromatography, which became established as the pre-eminent method for separating and analysing mixtures of volatile components. So valuable did it prove, that all manner of ingenuity was employed in extending its scope, both in respect of the samples which would be handled by it, and the things which could be done to them. This enormous expenditure of effort has meant that most worthwhile developments have been investigated and there have been few major changes since 1970; the subject has matured and settled down, and it is difficult to foresee major developments in the near future.

Like all chromatographic techniques, gas chromatography separates mixtures by taking advantage of their components' differential distribution between two phases – one stationary and the other moving past it. The distinctive feature of gas chromatography is the use of a gas as the moving, or mobile, phase. A sample of the mixture to be separated is introduced into this gas stream just before it encounters the stationary phase; the components are separated by elution and detected as they emerge in the gas at the other end of the column. They are distinguished by the different times which they take to pass through the column – the retention times.

The retention time of a substance is dictated by the position of its distribution equilibrium (see ACOL: *Chromatographic Separations*) between the two phases; separation of a mixture, therefore, depends on its components having significantly different distribution equilibria. Because there is little interaction between molecules in the gas phase, the gaseous mobile phase plays a mainly passive role in the separation, serving merely to carry the components through the system. The distribution equilibria are effectively controlled by the components' vapour pressures and their sorption by the stationary phase. Accordingly, the separation has to be carried out at a temperature at which the components' vapour pressures are high enough to allow a realistically short analysis time, but at which the differences between their vapour pressures is proportionately high. The stationary phase has to be chosen so that it forms a stronger attraction for one component than for another. It is, in practice, the stationary phase that often makes the greatest contribution to a separation, since interactions between it, as solvent, and the components, as solutes, are both strong and varied. Historically, the first stationary phases used were solid adsorbents; except in the field of permanent gas analysis they have been largely replaced by liquid stationary phases. These are more reliable and give more reproducible results in the analysis of volatile liquids. In fact the enormous growth in gas chromatography dates from the introduction of liquid stationary phases and the abbreviation 'glc' (standing for *gas–liquid chromatography*) is often inadvertently, and quite incorrectly, used when what is meant is the whole field of gas chromatography.

The range of liquid stationary phases available is enormous; specialist supply houses often list more than 200 in their catalogues. There is often much near duplication in such lists, but their size indicates the effort which has gone into finding the ideal stationary phase for a given task (and the difficulty of doing this!). One aspect which has received a great deal of attention is the decomposition and evaporation of stationary phases at high temperatures. This sets an obvious limit on the highest temperature at which they can be used which in turn means that glc can be used to separate only volatile or moderately volatile mixtures. One approach to overcoming this problem has been to search for stationary phases with ever lower vapour pressures and greater thermal stabilities. Such stationary phases can be used at higher temperatures, and so permit the analysis of less volatile samples by glc.

Π Suggest two classes of compounds from the following list which might be worth investigating for use as stationary phases at high temperatures:

Alkanes, tetraalkyl ammonium hydroxides, polyesters, polyethers, acid chlorides, silicones, acids, polyamides, salts, aqueous solutions of salts, amines.

1.

2.

If the compounds which you have suggested are of very high molar mass, you are on the right lines, as such compounds usually have low vapour pressures. Polymers, of course, are good examples, although polymers with fairly reactive functional groups, such as polyesters, will not stand up to the highest temperatures. In fact, silicone polymers and polyamides have been more valuable for very high temperature work, as have the very high molar mass alkanes found in petroleum greases. If you suggested thermally unstable compounds (tetraalkyl ammonium hydroxides), reactive compounds (acids, amines, acid chlorides), non-liquids (salts) or volatile combinations (aqueous solutions) perhaps you need to think a little more about the conditions to which the stationary phase is exposed and what it has to do.

Another approach to extending the range of samples for which glc can be used has been to increase the volatility of the sample. Since hydrogen bonding is often the cause of a sample's high boiling point, replacing the hydrogen atoms responsible by alkyl or silyl groups, often reduces the boiling point. This process, of turning samples into esters, ethers and trimethylsilyl derivatives, is called *derivatisation.* Where this approach is unsuitable, say in the case of a sample that is polymeric, an alternative has been to fragment the molecule by pyrolysis and to submit those fragments that are volatile to gas chromatography – the technique of *pyrolysis* gas chromatography.

A further problem which has had to be addressed is the difficulty of analysing very complex mixtures. Golay attacked this by dramatically increasing the efficiency of the column, thereby improving resolution and so separating these mixtures. He did this after careful consideration of the causes of peak broadening led him to the conclusion that irregularities in the packing were a major factor. He did away with the solid powders onto which the liquid stationary phases had been coated and deposited the liquid stationary phases directly onto the walls of very narrow bore tubes of considerable length. These have become known as *capillary columns*, or *open tubular columns*.

A further aspect of such complex mixtures is that their components may have a very wide range of boiling points, so that no one analysis temperature would be suitable for all of them. Here the solution has been to start the analysis at a low temperature, appropriate to the most volatile component, and to increase it during the analysis until the least volatile component has been eluted at an appropriately high temperature. This technique has been christened *temperature programming*.

Complex mixtures may also contain different types of components (eg chlorinated and non-chlorinated hydrocarbons), of which only one class is of interest. A solution to this has been to seek to develop selective detectors which respond only to the class of compounds that is of interest. It is not then necessary to achieve complete separation of the mixture since those components to which the detector does not respond cannot cause interference. Such detectors also have some value in the context of another problem encountered in gas chromatography. Gc may be an excellent technique for analysing a mixture when you know where it has come from and what it is likely to contain, but when you are *starting from cold* with a mixture about which you know nothing, a retention time (which is all that gas chromatography gives you) does not tell you a great deal. At least a *selective detector* might tell you what class of compound you are looking at. Ideally, of course, you want the sort of structural information provided by molecular spectrometry if you are to identify the separated components, and this has lead to the development of such techniques as combined gc/ir and gc/ms. Early methods of combining gc with molecular spectrometry involved trapping the separated components as they emerged from the column, usually in cold traps, and then transferring them to a spectrometer. This was tedious and inefficient, and improvements in spectrometers and in the techniques for interfacing them with gas chromatographs now allow the gas stream from a gas chromatograph to be led directly into a spectrometer and for the spectra to be recorded *on the fly*, ie without stopping the gas flow. Of the techniques mentioned above, gc/ms has been by far the most successful combination although gc/ir is gaining ground rapidly since the introduction of a new generation of Fourier transform spectrometers. In conjunction with computer – based data handling and libraries of spectra, both are very powerful, if expensive, techniques.

The amount of material which could be obtained from a conventional gas chromatograph by the trapping technique described above was barely enough, and larger columns, which could cope with bigger samples, were developed. By employing repeated injections of the sample and carefully timed trapping of separated components under automatic control, so that they were *bulked up*, modest amounts of material could be obtained and gas chromatography developed from an analytical into a preparative technique.

Π Now that you have read this introduction, try defining gas chromatography:

Gas Chromatography is ...

If your definition refers to gc as a technique for separating mixtures which relies upon the differential distribution of their components between a stationary, condensed phase and a mobile, gaseous one, you have been thinking along the right lines. If you referred only to analysing mixtures or to liquid samples, you were being too specific.

Gc is a very wide ranging technique, as I have hinted at above, and as you will discover as you read on....

Summary

The history of gas chromatography has been traced from its introduction as a very simple technique in the 1940's to its present highly developed form. It has been described as a technique capable of separating volatile mixtures for analytical or preparative purposes and the basis of its ability to separate mixtures has been discussed. The many valuable modifications and adaptations of the technique have been introduced.

SAQ 1.a

Indicate, by circling either T for True or F for False, whether you agree with each of the following statements:

(*i*) The best way to analyse a mixture of H_2O and D_2O is by glc.

T / F

(*ii*) If a mixture of ethane and ethene is not separating very well with argon as the carrier gas, it can be improved by switching to hydrogen.

T / F

(*iii*) If you want to analyse a mixture of 1,2-, 1,3- and 1,4-dimethylbenzenes you are likely to have to use a capillary column.

T / F

(*iv*) If you wanted to identify the contents of an unlabelled bottle containing a mixture of organic solvents that has been found on the street, you would use gc.

T / F

SAQ 1.b

Indicate, in the space provided, the technique, chosen from the list A to E, that would be appropriate for:

(*i*) Distinguishing between samples of polyethylene and polypropylene

(*ii*) Analysing a mixture of phenols

(*iii*) Estimating polychlorinated biphenyls (PCB's) in infertile sparrowhawks' eggs

(*iv*) Analysing a mixture of petrol and diesel fuel

A = pyrolysis gas chromatography;
B = use of a selective detector;
C = derivatisation;
D = temperature programming;
E = preparative gc.

Learning Objectives

After studying the material in Part 1, you should now be able to:

- define gc and delineate its boundaries;
- discuss the type of sample which could be analysed by gc;
- discuss the various ways in which samples can be analysed by gc.

2. The Working Gas Chromatograph

2.1. FUNDAMENTALS

The design of the instrument that you would need to build in order to implement in practice the principles described in the Introduction follows from the decision to use a gas as the mobile phase. The basic concept of such an instrument has remained unchanged since the first one was built, although there has been much refinement and an amazing improvement in performance. The main components are shown in Fig. 2.1.

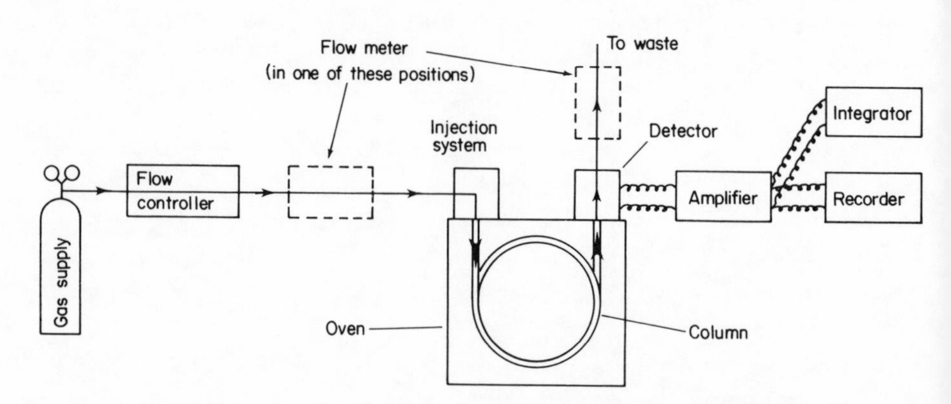

Fig. 2.1. *Block diagram of a gas chromatograph*

All gas chromatographs must have a supply of gas to act as mobile phase and some means of controlling and measuring its flow rate. They must have a column, which is normally contained within a thermostatted oven. There will be some means of introducing the sample into the gas stream, usually called an injection system, and some means of detecting the separated components as they emerge from the column. The detector will have ancillary equipment associated with it to provide a permanent record of the analysis and to manipulate the data.

2.2. THE CARRIER GAS

The mobile phase (often called the carrier gas) most frequently used in the UK is nitrogen. The second most popular is helium, which is more expensive, but essential for certain applications; argon, hydrogen and carbon dioxide have also been used, though much less frequently. Although the physical properties of the carrier gas can influence the quality of a separation, this effect is small, and compatibility with the detector is the factor that usually controls the choice of gas.

The gases can be obtained from commercial suppliers, conveniently compressed in cylinders and in a state of purity sufficient for most purposes. If, however, the gases are impure, the high standing signal produced by the detector responding to the impurities fluctuates with each change in conditions and a very noisy baseline results. The commonest impurity is water, and drying the gas with an in-line molecular sieve trap is normally recommended. The other impurity which might give trouble is oxygen, so *white spot* nitrogen, which has a very low oxygen content, is used.

Π Why is oxygen in the carrier gas bad news?

Why is this especially so when you are working at high temperatures?

If you suggested that it would cause oxidation of the stationary phase, you were thinking along the right lines. The stationary phases are usually organic, and the products of oxidation are usually volatile and cause a noisy, drifting baseline when they reach the detector. The column will also deteriorate, so that after a while it will no longer give such good results (its life is shortened). This oxidation is more rapid at high temperatures, of course, and more of the products of oxidation reach the detector.

The gas needs to be conveyed, usually at a pressure of about 40 psi, or 3 bar, to the chromatograph. Conventional cylinder-head pressure reducers and 1/8 inch diameter nylon or metal pipework are sufficient for this, but they do not provide fine enough control of the flow rate on their own. It is usual to include a good quality needle valve (Fig. 2.2a) in the supply line to do this. A needle valve works by advancing a tapered needle into a conical hole so that it constricts the gas flow to the required valve. With enough care, the valve can be made smooth and progressive in its action, but obviously a change in room temperature, or a change of back pressure in the column will affect the flow rate.

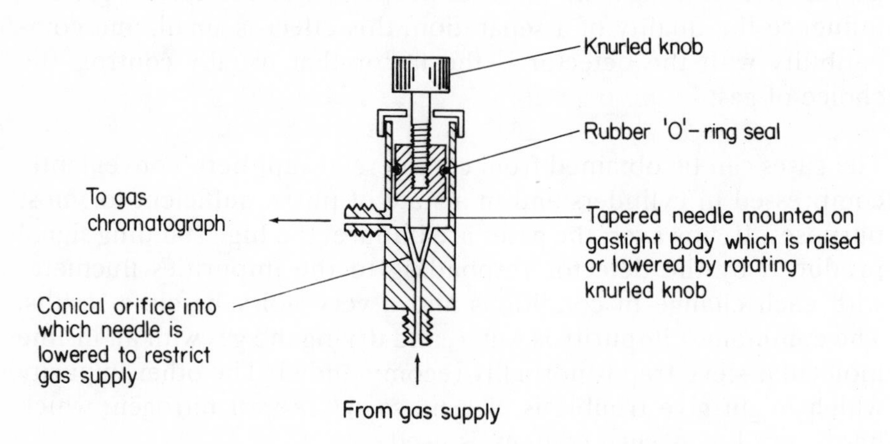

Fig. 2.2a. *Needle valve or flow controller*

Measuring the flow rate is another problem altogether. None of the flow meters commonly used is completely satisfactory, since they all ought to be recalibrated for each change in ambient temperature and pressure. Of those which can be connected in-line, before the gas reaches the column, the type which measures how high a ball is carried by the carrier gas up a narrow, tapering glass tube (Fig. 2.2b) is more popular than the type which measures the pressure drop across a capillary (Fig. 2.2c).

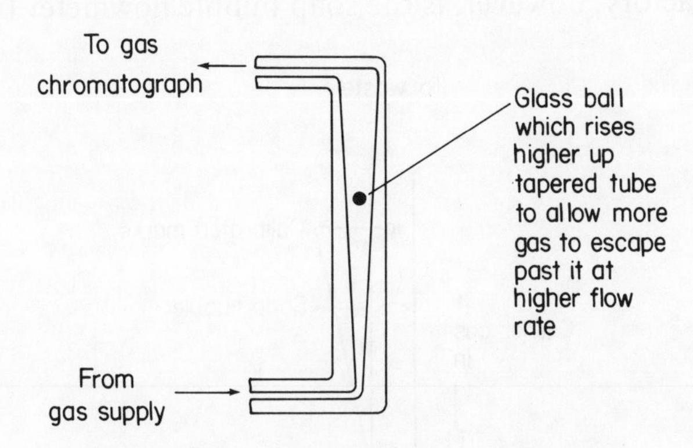

Fig. 2.2b. *Mobile ball flow meter*

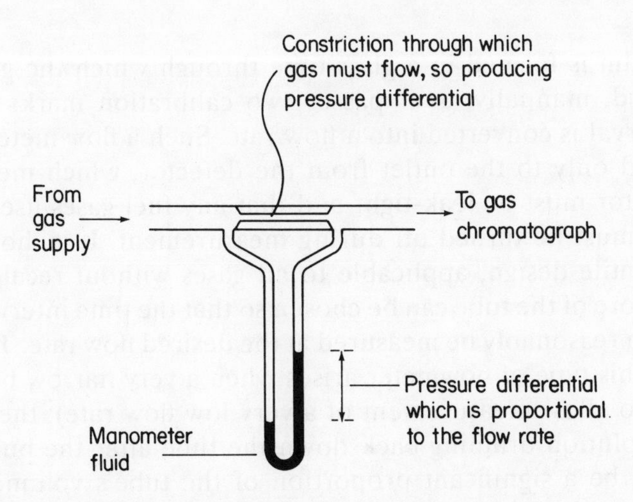

Fig. 2.2c. *Manometric flow meter*

Each of them needs to be calibrated for the gas in use, since the reading will depend upon the viscosity and density of the gas, but the former is more easily designed as a direct reading instrument covering a wider range of flow rates and does not suffer from the risk of inadvertent migration of the manometer fluid! Probably the most satisfactory, however, is the soap bubble flowmeter (Fig. 2.2d).

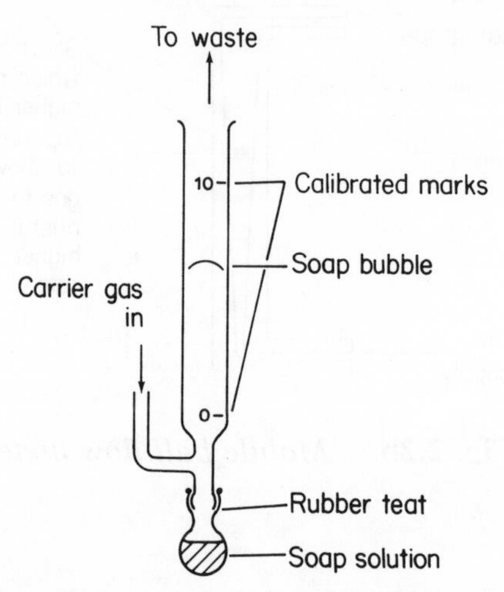

Fig. 2.2d. *Soap bubble flow meter*

A soap film is formed in a glass tube through which the gas flows. It is timed, manually, as it passes two calibration marks and the time interval is converted into a flow rate. Such a flow meter can be connected only to the outlet from the detector, which means that the detector must be leak-tight and that any fuel gases used by the detector must be turned off during measurement. It is, however, a very versatile design, applicable to all gases without recalibration, and the bore of the tube can be chosen so that the time interval is one which can reasonably be measured at the desired flow rate. The main error of this type of flowmeter arises when a very narrow bore tube is used (to allow measurement of a very low flow rate); the volume of soap solution draining back down the tube after the bubble has burst can be a significant proportion of the tube's volume, which means that the volume of gas assumed to pass when the next bubble is timed can be in error.

2.3. THE OVEN

The column of the gas chromatograph must be encased in a well thermostatted oven because gas chromatography is much more sensitive to changes in temperature than liquid chromatography. This is not, as you might suppose, because temperature changes affect the flow rate significantly, with all that this would mean for changed retention times and zone spreading (see ACOL: *Chromatographic Separations*, Parts 2 and 3). It is true that this might happen in liquid chromatography, for you can see from Fig. 2.3a that a 100 °C rise in temperature reduces the viscosity of benzene by about one third. This would cause the flow rate to increase by about the same amount, if benzene were being used as a mobile phase, as the resistance to the liquid's flow through the column would fall.

	Nitrogen (gas)		Benzene (liquid)	
	27 °C	127 °C	27 °C	127 °C
Viscosity (millipoise)	0.178	0.219	570	330
Volume/Volume at 27 °C	1.00	1.33	1.00	1.05

Fig. 2.3a. *Viscosities of nitrogen and benzene*

However, in gas chromatography, the actual speed of the gas within the column (the discussion in *Chromatographic Separations* of this series made it clear that this is what is important) stays much more nearly constant. Although a gas expands much more than a liquid when it is heated up on entering the column, so that the bigger volume of gas has to move faster to get through the column, its viscosity actually increases as the temperature rises (Yes! I always find that surprising too!). This will have the effect of increasing the resistance to the flow of the gas through the column and reducing its speed. The two effects almost balance out and the speed of the gas hardly changes when the temperature changes. A word of warning, though. The flow rate which you would get by measuring it with a soap bubble flow meter would change. By the time the gas reached the flow meter, it would have cooled down to room temperature and contracted again.

What is much more significant is the effect of temperature changes on the position of the distribution equilibrium. In liquid chromatography, an increase in temperature will increase the solubility of components in both phases, so that the distribution equilibrium does not change overmuch. However, in gas chromatography, an increase in temperature will usually increase the vapour pressure of a component by much more than its solubility in the stationary phase.

	20 °C	70 °C
Solubility of water in benzene (g dm^{-3})	0.57	2.7
Saturated vapour pressure of water (mm of Hg)	17	233

Fig. 2.3b. *Effect of temperature on vapour pressure and solubility*

This causes a large disturbance in the distribution equilibrium position and therefore temperature control assumes a far greater importance in gas chromatography.

The ovens used are generally small, approximately cubic, and with forced air circulation. Sensitive thermocouples allow accurate thermostatting, and one example claims to hold the temperature constant to ± 0.05 °C, to be capable of heating up from room temperature to 200 °C in 10 minutes and cooling back down again to 50 °C in 17 minutes. There has been some debate over the merits of a lightly insulated, low thermal mass oven over a well insulated one like the one referred to above. It would offer much less stability, but it would be capable of more rapid heating and cooling. This can be important in the technique of temperature programming, where a repeat analysis cannot be conducted until the oven has cooled back down to the start temperature after the previous analysis – 17 minutes can seem an awful long time! The argument has not really been settled, although it is true to say that most commercial instruments err on the side of good lagging.

2.4. THE COLUMN

The focus of all this ancillary equipment, the chromatographic column, is enclosed in the oven. Traditionally, it was a glass tube, 4 mm in diameter and about 1.5 mm long, coiled so that it fitted into a small oven which could be effectively thermostatted. Such columns were then filled (*packed*) with a finely divided solid (the *packing*) which might be either a solid adsorbent, or else an inert solid (the *support*) which had been coated with a thin film of a non-volatile liquid (the *liquid stationary phase*). The volume of such a column is not large, and the amount of stationary phase which it can contain will be limited (about 10 g of a solid adsorbent or about 1 g of a liquid stationary phase). It will not be able to handle a very large sample.

Π Estimate the length of the 'plug' of vapour formed by 0.1 g of benzene if it is injected onto a column 1.5 m long and 4 mm id, packed with a support that has been coated with a liquid stationary phase, at a temperature of 80 °C. The particles with which the column is packed will occupy about two thirds of the volume of the column, leaving spaces between for the gas to percolate through which will account for about one third of the volume of the column. (The relative molecular mass of benzene is 78).

Comment on the implications of this for its separation from any other liquids that might be mixed with it.

If you came up with the answer that it is nearly as long as, or longer than the column, don't be too surprised. You are in the right region. I worked it out like this:

Volume of the benzene vapour at STP

$$= \frac{0.1 \times 22}{78} \text{ dm}^3$$

Volume at 80 °C

$$= \frac{353 \times 0.1 \times 22}{300 \times 78} \text{ dm}^3$$

$$= 33 \text{ cm}^3$$

The cross sectional area of column

$$= \pi \times 0.2 \times 0.2 \text{ cm}^3$$

But only 1/3 of this is available to the gas phase,

Available cross sectional area

$$= \frac{\pi \times 0.2 \times 0.2}{3} \text{ cm}^3$$

$$= 0.04 \text{ cm}^3$$

33 cm of vapour would make a plug

$$= \frac{33}{0.04} \text{ cm}^3$$

$$= 825 \text{ cm long!}$$

The column is only 150 cm long, so it would be quite a broad band!

Π Have you spotted the omission in the argument?

I have assumed that all the vapour is in the gas phase, when of course some will be dissolved in the liquid stationary phase. At equilibrium, probably about 90% of it, if benzene is to have a reasonable retention time. That still leaves enough vapour to make a plug 82.5 cm long, which is over half the length of the column. Separation would therefore be pretty poor.

2.5. THE INJECTION SYSTEM

Samples are usually less than 1 mg (eg 1 μl of liquid or 5 cm^3 of gas). Such small samples require special handling techniques, both for injecting them onto the column and monitoring them as they emerge from the column. Injection, is usually achieved by means of a syringe inserted through a self-sealing silicone-rubber septum (Figs. 2.5a and 2.5b)

The larger syringe (*i*), for gases, is perfectly conventional, although great care is taken during manufacture to ensure that the plunger is a very good fit in the barrel. If it is not a good fit, pressure inside the column tends to blow the sample back past the plunger. For moderate quantities of liquids, a very similar design, albeit with much narrower bore, is used, although the plunger is almost always of stainless steel. As the volume of the syringe gets smaller, though, the *dead volume* in the needle becomes appreciable and at a volume of around 1 μl an alternative design (*ii*) is used. The plunger is a wire of the same diameter as the bore of the needle and is extended right down to the tip of the needle, so that there is no dead volume. Because of the frailty of this wire, it has to be supported by a series of concentric tubes which slide within one another and prevent (hopefully!) the wire from kinking as the plunger is pushed home. It works, provided you are very careful, but stiff or leaking microsyringes are the bain of a gas chromatographer's life. There are a number of problems inherent in the use of syringes for injection, even when they are not damaged:

(*i*) Even the best syringes claim an accuracy of only 3%, and in unskilled hands, errors are much larger.

(*ii*) The needle cuts small pieces of rubber from the septum as it penetrates it. These can block the needle and prevent the syringe filling the next time it is used, without it being obvious what has happened.

(*iii*) A fraction of the sample may get trapped in the rubber, to be released during subsequent injections or increases in temperature. This can give rise to artefacts in a later analysis known as *ghost peaks*.

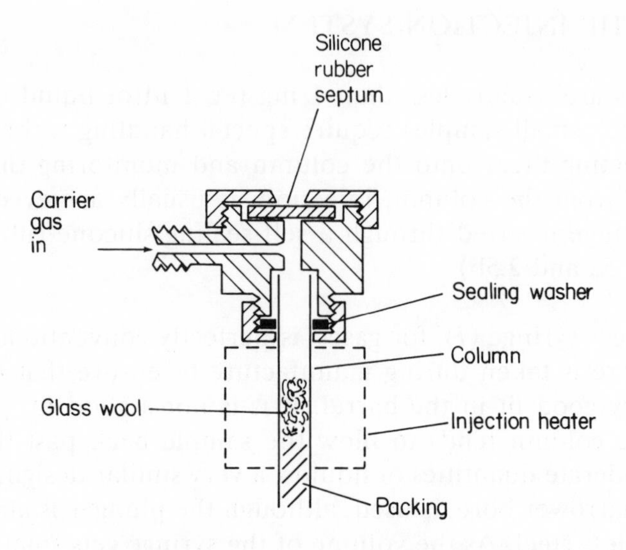

Fig. 2.5a. *Injection Head*

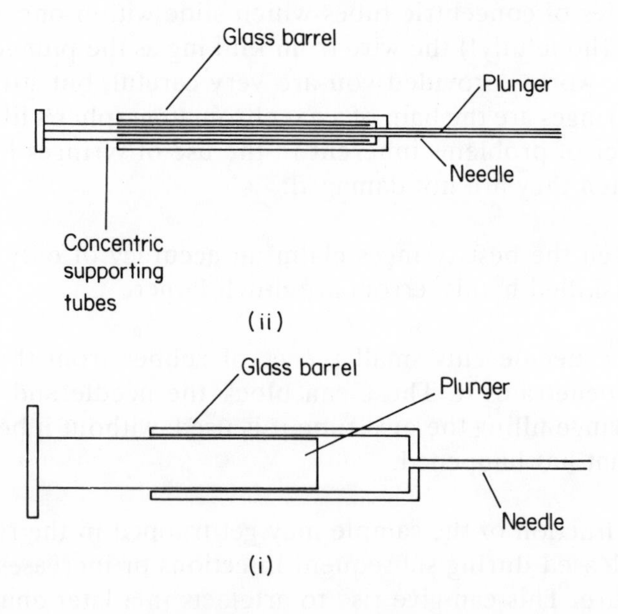

Fig. 2.5b. *Syringes*

(*iv*) If there is a delay between filling the syringe and injecting the sample, there may be selective loss of the more volatile components by evaporation from the tip of the needle.

(*v*) The minimum sample that can be delivered in this way is 0.1 μl.

At this level the accuracy of delivery will be poor, so an internal standard (see ACOL: *Chromatographic Separations*) must be used for quantitative work. Furthermore 0.1 μl would overload a capillary column (see Part 3 of this Unit). This latter problem can be overcome by the use of a stream splitter (Fig. 2.5c):

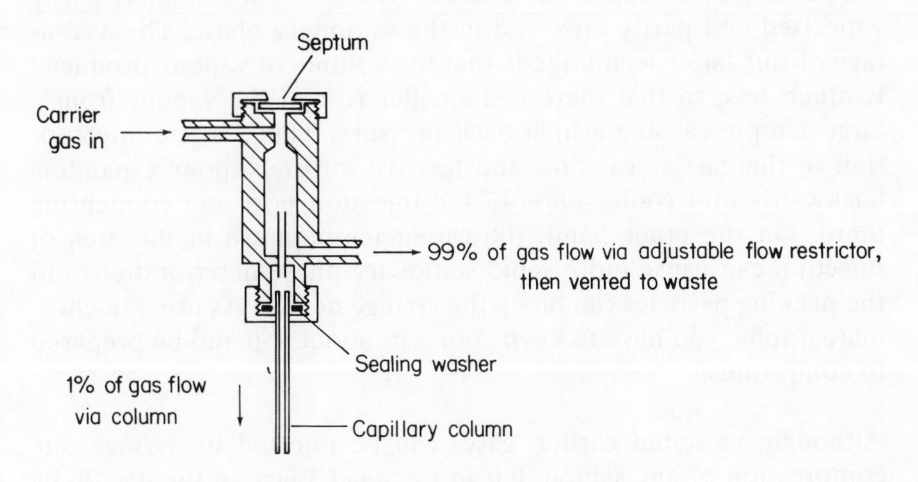

Fig. 2.5c. *Stream splitter*

On injection, the sample is vaporized, diluted with carrier gas, and a fraction of this mixture delivered to the column, the remainder being vented to the atmosphere. A major difficulty for quantitative work is that the split may not be homogenous. This may be because there is poor mixing with the carrier gas at the dilution stage or because the low relative molecular mass components diffuse towards the vent more rapidly than those of higher relative molecular mass. The splitter is said to show *discrimination*.

No matter how the liquid is being injected, it is obvious that it must be vaporized quickly, so that it enters the carrier gas stream as a narrow band. This is usually done by surrounding the region of the column which the tip of the needle reaches with an electrical heater. The raised temperature ensures rapid evaporation, but it can also cause decomposition of thermally sensitive samples, so a little care is needed in choosing the temperature at which to set the heater. Another variation is to use either *flash vaporisation* or *on-column* injection. In flash vaporisation, the sample is injected either into the unpacked space at the top of the column or into an empty space incorporated into the injection port, which may be fitted with a replaceable glass liner. In both cases, the sample is vaporised in the empty space by the injection heater and then swept onto the column. In on-column injection, the sample is injected directly into the packing at the top of the column, where it is immediately partly vaporised and partly dissolved in the stationary phase. The advantage of this latter technique is that the volume of vapour produced is much less, so that there is a smaller risk of the vapour from a large sample causing a high back-pressure. There is less interruption to the carrier gas flow and less risk of the vapour expanding backwards into cooler parts of the injection port and condensing there. On the other hand, the raised temperature in the area of injection can cause more rapid stationary phase deterioration, and the packing particles can block the syringe needle. As usual in chromatography, you have to keep your wits about you and be prepared to compromise.

Although, as stated earlier, gases can be injected by syringe, the compression of any sample left in the dead space in the needle by the pressure within the column can lead to inaccuracies which may be unacceptable in quantitative work. In this case, a multiport valve and sample loop (Fig. 2.5d) offer advantages.

With the rotor of the multiport valve in position (*i*), the calibrated sample loop can be filled with the sample gas at atmospheric pressure, simply by flushing it. The rotor can then be turned to position (*ii*) and the sample will be carried straight onto the column by the diverted carrier gas stream.

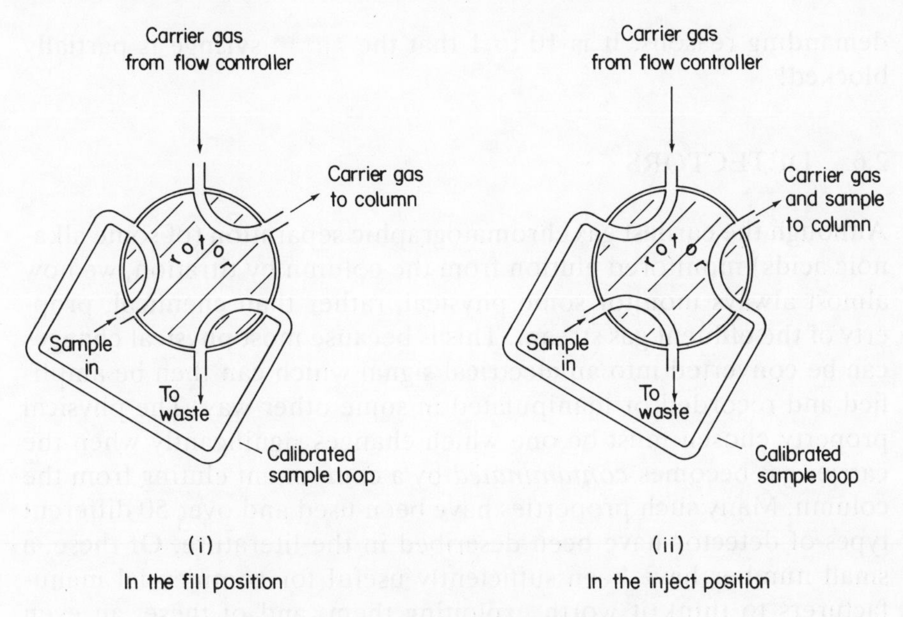

Fig. 2.5d. *Sample loop for gases*

Quite apart from their use for the analysis of gases as such, syringes and sample loops are also used for *headspace analysis*. In a sealed container partially filled with a solution of volatile components, the composition of the mixture of vapours above the solution will be such that it is in equilibrium with the solution. If the container is carefully thermostatted, analysing this mixture of vapours will allow the composition of the solution to be determined. This is the technique of 'headspace analysis'. After proper calibration, it can offer accuracy comparable to that obtained by conventional analysis of the liquid phase and has the advantages that sampling is more easily automated and the column is not contaminated by the non-volatile components of the solution.

You will probably have gathered from this discussion that injection of the sample is one of the critical areas of gc. It probably gives more trouble, especially to beginners, than any other single area. I make no apology for stressing so mundane a point as the need for the proper care of syringes, valves and septa or for suggesting that if you are having difficulty with an analysis you look first at your injection technique before you start looking for more intellectually demanding reasons. It is 10 to 1 that the ***** syringe is partially blocked!

2.6. DETECTORS

Although the earliest gas chromatographic separation (of some alkanoic acids) monitored elution from the column by titration, we now almost always monitor some physical, rather than chemical, property of the effluent gas stream. This is because most physical changes can be converted into an electrical signal which can then be amplified and recorded or manipulated in some other way. The physical property chosen must be one which changes significantly when the carrier gas becomes *contaminated* by a component eluting from the column. Many such properties have been used and over 50 different types of detector have been described in the literature. Of these, a small number have been sufficiently useful for commercial manufacturers to think it worth exploiting them, and of these, an even smaller number remain in common use. If you ask yourself why this should be, you will find that you are trying to answer the question 'What properties would I look for in a good detector?' In fact, you might like to do that before you continue (I can think of about seven).

Π List the properties which you think a gas chromatography detector should have (don't worry if you cannot think of seven – suggest as many as you can).

My list, in approximate order of importance, is as follows:

Sensitivity, Stability, Linearity, Universality, Selectivity, Ease of use, Cost,

(*a*) Sensitivity is usually defined as the response per unit concentration of analyte. Although you will often see the sensitivity of a detector quoted as '10^{-3} moles' or '10^{-y} moles%', it ought to be in 'mV mg^{-1} cm^{3}'.

Sensitivity can be calculated as:

$$\text{sensitivity (mV mg}^{-1}\text{ cm}^3) = \frac{\text{A} \times \text{FR}}{\text{RS} \times \text{CS} \times \text{W}}$$

Where A = peak area (cm^2)
RS = recorder sensitivity (cm mV^{-1})
CS = chart speed (cm min^{-1})
FR = flow rate of carrier gas (cm^3 min^{-1})
W = weight of sample injected (mg)

The sensitivity determines the slope of the calibration graph, and so, to some extent, the precision of analysis. Although it is not the same thing as limit of detection, the two are closely related, and a high sensitivity often means a low limit of detection, which means that you will be able to detect and determine very small quantities of analyte. You must remember, however, that the limit of detection is usually defined as *that concentration or amount of analyte which produces a signal equal to twice the baseline noise* (see stability, below, for a definition of noise), so that a detector with a high sensitivity will not have a low limit of detection if it is also subject to excessive noise.

(*b*) Stability is the extent to which the output signal remains constant with time, given a constant input. Instability can take two forms, a rapid, random variation in output, or *noise*, or a slow, systematic variation, or *drift*. Both of them will limit the sensitivity of the detector; noise because it will be difficult to pick out small peaks against a noisy background and drift because the baseline will be disappearing off scale during the analysis. Fig. 4.6 shows something of this. Imagine how difficult it would be to cope with a baseline which was sloping like the one at Attenuation ×1, or to measure the area of peaks only a few millimeters high on such a baseline.

(*c*) Linearity, is the extent of the range over which the signal is truly proportional to the concentration or amount of analyte.

In effect, linearity refers to the linearity of the calibration graph. A calibration graph can be drawn with more confidence if it is a

Barton College Library
Wilson, N.C. 27893

straight line. With a convex calibration curve (Fig. 2.6a), and this is probably the most common type, the precision will be much poorer at high concentrations, where the slope has diminished, than at low concentrations, where the slope is relatively steep, and there will come a point at which the method is no longer viable because the precision is so bad that the results are almost meaningless.

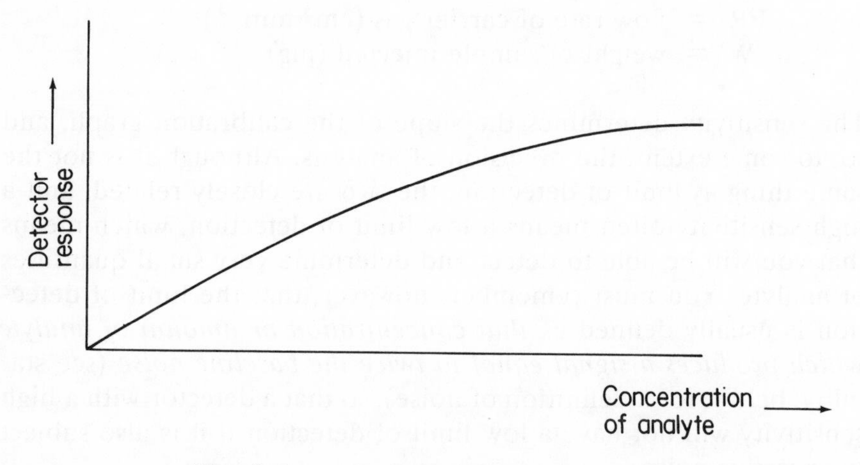

Fig. 2.6a. *Convex calibration graph*

The linearity of a detector is usually described by quoting its linear dynamic range, or the range over which its response is essentially linear, expressed as the factor by which the lowest concentration must be multiplied, to obtain the highest concentration. We shall see that this can be very high (10^6 for some detectors), and this is a great advantage. For other detectors a very small linear dynamic range is tolerated because of their other qualities, and this will almost certainly mean that they will need to be carefully recalibrated over each of a number of concentration ranges.

(*d*) Universality is the detector's ability to detect all the components present in a mixture. If it is not able to detect them all, or if the sensitivity for different components is markedly different, then there is an element of uncertainty in any analysis undertaken and an obvious limit to the application of the detector.

(*e*) Selectivity is the opposite of universality (*c*). If it can be arranged so that a detector *ignores* unwanted components in a mixture and only responds to those which are of interest, the resulting chromatogram may be considerably simplified. Interferences could, in this way, be avoided and it might even be possible to eliminate an otherwise necessary *clean up* stage on the sample before gas chromatography is undertaken.

You can see how the selectivity of a detector can be used to simplify an analysis by comparing the chromatograms shown in Figs. 2.6f and 2.6g. They were obtained by analysing the same mixture using two different detectors, one of which was much more selective than the other for the two pesticides present.

(*f*) Ease of use is a criterion which is almost too obvious to need stating, but some detectors are more tricky to *set up* than others and may require more frequent maintenance, or the maintenance may be more difficult. Other things being equal, this could tip the balance in the choice between two detectors, although the ability to do the job, as judged by (*a*) to (*d*) must remain the first criterion.

(*g*) Cost will also be a factor which must be taken into account. Whilst capital cost is important, it may be that running costs, in the form of the price of any special gases needed, the cost of maintenance or of *down time* in the case of less robust equipment, may be of even greater significance.

In this course I intend to discuss three commonly used detectors with you under the headings referred to above. I shall make some reference to other commonly used detectors where this is pertinent to the discussion. Should you want to read more about them, I recommend that you consult Ettre and Zlatkis' book which is one of the most comprehensive discussions of detectors I have come across.

2.6.1. The Katharometer or Thermal Conductivity Detector (TCD)

This was the earliest successful detector and is still in use today as a good, general purpose detector. It was, in fact, an adaptation of a device already in use as a gas analyser or monitor, and it is its simplicity and robustness, together with its ability to detect almost everything, which has lead to its continued use. It usually takes the form of a thermostatted metal block with two passageways drilled in it. Each passageway has fitted into it a wire, or filament, made from a metal with a high temperature coefficient of resistance, such as platinum or tungsten. The two filaments are connected into the arms of a Wheatstone bridge. Pure carrier gas is made to flow through one passageway (the reference cell) and the effluent from the column through the other (the analysis cell).

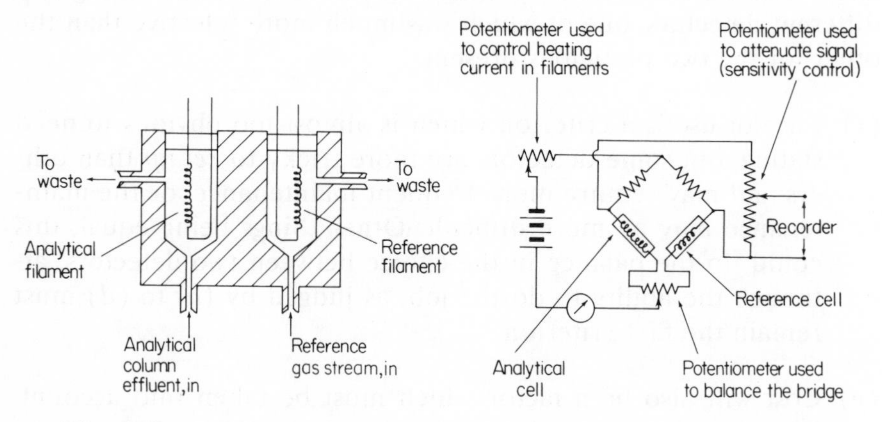

Fig. 2.6b. *Schematic diagram of katharometer and circuit*

The principle upon which the katharometer depends for its operation is that the rate of heat loss from a heated wire placed in a gas stream depends on, amongst other things, the thermal conductivity of the gas. Thus, if the composition of the gas stream changes, and hence its thermal conductivity, the rate of heat loss from the wire will change and the wire will settle to a different equilibrium temperature and a different electrical resistance. Since two similar filaments are placed in a balanced Wheatstone bridge with pure carrier gas flowing over one of them and the effluent gas from the chromatography column over the other, any change in the composition of the column effluent due to the elution of a component leads to a change in its thermal conductivity which throws the bridge out of balance. The out-of-balance potential, which can be recorded, will be related to the concentration of the component in the effluent stream. A graphical record of the output from the Wheatstone bridge will take the form of a steady baseline with *peaks* superimposed on it, each peak occurring with its maximum at a position corresponding to the retention time of a component and having an area related to the amount of component that was injected.

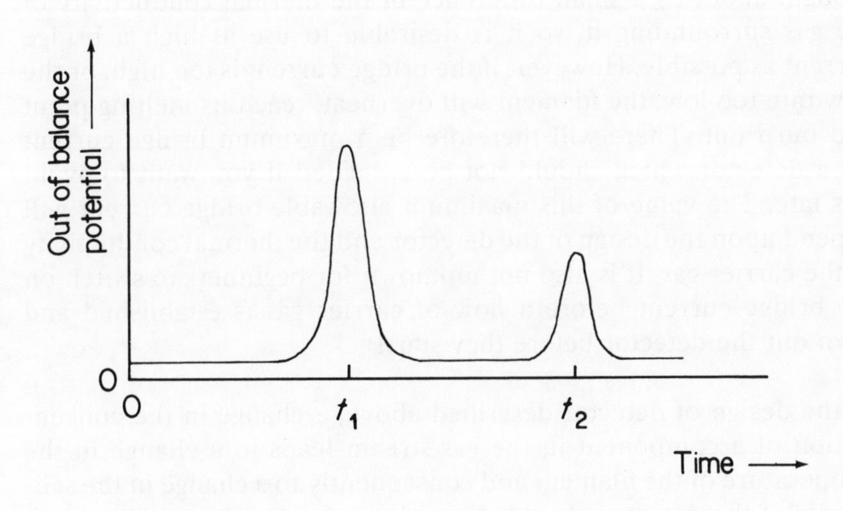

Fig. 2.6c. *A typical chromatogram*

Of course, fluctuations in any other condition which alters the rate of heat loss from the wire will lead to distortions of the baseline and changes in the sensitivity of the detector, and steps are normally taken at least to minimise these effects.

Π See if you can suggest three parameters which will have a major effect upon the output from a katharometer.

If you have suggested the temperature of the wire, the temperature of its surroundings and flow rate you have chosen the three important ones. You may have included the nature of the carrier gas, since its thermal conductivity will have an obvious effect, but I think that is a bit like shooting fish in a barrel!

Other things being equal, the hotter the wire, the greater the rate of heat loss from it; since the wire is being heated by the electric current flowing in the Wheatstone bridge, it follows that the circuitry employed must maintain the current as nearly constant as possible if the detector is to be stable. Furthermore, the higher the temperature of the wire, the greater will be the difference in rate of heat loss brought about by a small difference in the thermal conductivity of the gas surrounding it, so it is desirable to use as high a bridge current as possible. However, if the bridge current is too high, or the flow rate too low, the filament will overheat, reach its melting point and burn out. There will therefore be a maximum bridge current for a detector which should not be exceeded if you want to avoid this fate. The value of this maximum allowable bridge current will depend upon the design of the detector and the thermal conductivity of the carrier gas. It is also not unknown for beginners to switch on the bridge current before a flow of carrier gas is established and burn out the detector before they start!

In the design of detector described above, a change in the concentration of a component in the gas stream leads to a change in the temperature of the filament and consequently to a change in the sensitivity of the detector. As this limits linearity, an alternative design has been developed. This employs the same physical arrangement, but uses a circuit which varies the current in the bridge so as to keep the filament in the analysis cell at a constant temperature regardless of the rate of heat loss from it, and then records the current which had to be supplied to achieve this. In this way, the sensitivity remains more nearly constant, linearity is improved, and as an added bonus, it should be impossible to burn out the filaments.

The temperature at which the detector is maintained will have an important effect on the rate of heat loss from the wire, and katharometers are carefully thermostatted, often in a separate *detector oven*, so that the column temperature can be altered without altering the sensitivity of the detector.

The carrier gas flow rate will also affect the rate of heat loss from the wire, and so the flow rate has to be is maintained as steady as possible. Nonetheless, small fluctuations are frequently observed and these limit the stability of this detector.

The sensitivity of a conventional katharometer is not great, say around 100 mV mg^{-1} cm^3, but it is sufficient for normal as opposed to trace analyses with packed columns. It is not really adequate for use with capillary columns, and for these it is necessary to use a miniaturised cell in which a thermistor (a semiconductor whose resistance changes sharply with temperature) replaces the wire filament. This is still barely satisfactory, especially as a thermistor should not be used above 100 °C which places a severe limitation on its use. The sensitivity increases with increasing filament temperature (bridge current) and also with decreasing temperature of the surrounding metal block. Sensitivity depends upon the difference between the thermal conductivity of the carrier gas and that of the component being detected, and it will therefore not be the same for all the components in a mixture. This rules out the area normalisation method of quantitation. Since most organic vapours have comparatively low thermal conductivities, sensitivity is highest when gases which have high thermal conductivities, such as hydrogen and helium, are used as carrier gas and lowest when nitrogen or argon, which have lower thermal conductivities, are used. The ratio of sensitivities using hydrogen, helium, nitrogen and argon as carrier gas is approximately 7 : 6 : 1 : 1.

The linearity of the conventional katharometer is greatest if there is a large difference between the thermal conductivities of carrier gas and analyte, that is when hydrogen or helium are being used as carrier gas. The linear dynamic range is then of the order of 10^3, which is not large, but is adequate for most purposes. Linearity is often poor because the relation between the composition of a mixture of analyte and carrier gas and its thermal conductivity is not linear, and in many cases this can lead to W-shaped peaks (Fig. 2.6d) being observed when you are working with the detector at its most sensitive.

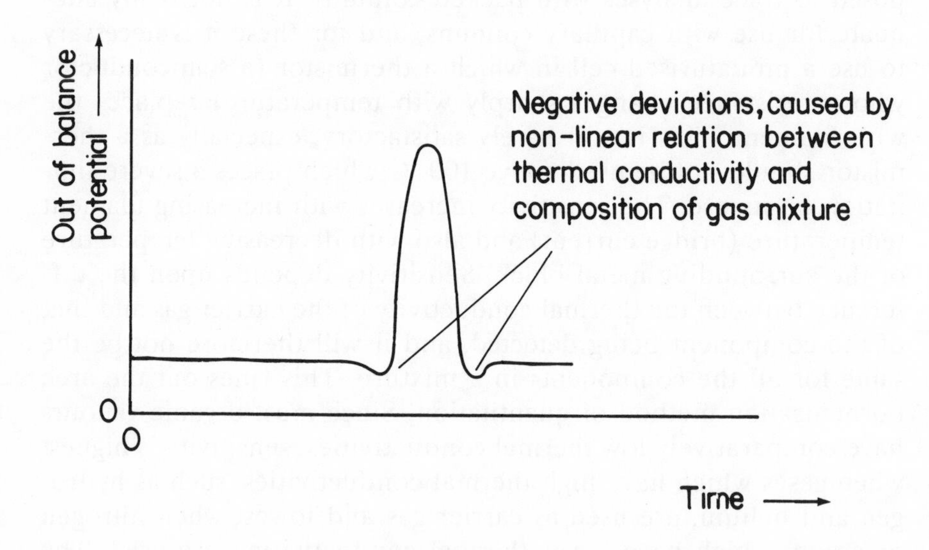

Fig. 2.6d. *W-shaped peak*

The katharometer is capable of detecting almost everything (has good universality), though the sensitivity for some analytes may be low if they have thermal conductivities close to that of the carrier gas. It is one of the few detectors which can be used to detect the inorganic gases and this means that it is the detector of choice for gas analysis. Conversely, its selectivity is poor although you can sometimes achieve the same effect of simplifying an analysis by using a component which you do not want to detect as the carrier gas. The atmosphere used for some forms of arc welding consists chiefly of argon and needs to be analysed for oxygen. Argon and oxygen are almost impossible to separate by gas chromatography, so the analysis has to be performed using argon as carrier gas, so that only the peak for oxygen is seen. The detector is unable to distinguish between the argon from the sample and the argon used as carrier gas.

The katharometer is robust and easy to use. It is relatively cheap to buy and to run, provided you do not decide to use helium as carrier gas in order to get maximum sensitivity. Although hydrogen is a lot cheaper and gives you even more sensitivity, it is not really a satisfactory alternative to helium because of its hazardous nature.

Π Which carrier gas would you choose for each of the following analyses, assuming that you have only a gas chromatograph fitted with a katharometer available to you?

(*a*) the determination of traces of hydrogen in air.

(*b*) the analysis of a mixture of propanone and propan-2-ol.

(*c*) the determination of argon in air.

(*a*) Nitrogen or argon would be the logical answer. Helium would not do because hydrogen also has a high thermal conductivity and so the sensitivity would be poor.

(*b*) Nitrogen would be satisfactory. Since, by implication, you do not need the ultimate in sensitivity (trace analysis is not specified), it would be extravagant to use helium.

(*c*) You didn't have a lot of choice. Since argon and oxygen are so difficult to separate, you have to use oxygen as carrier gas in order to eliminate interference from oxygen in the air sample. But you had better be careful. Use the lowest bridge current you can get away with in order to minimise the risk of oxidising your filaments. You will probably not have to worry about oxidising your stationary phase since you will almost certainly be using an inorganic molecular sieve column (see section 3.3 of this Unit) which would not be prone to oxidation.

Summary

The katharometer is a simple, robust, cheap to buy and cheap to run detector. Its sensitivity and linearity are only moderate –good enough for routine analysis but not for trace analysis. It is a good general workhorse of a detector which can be used for the general analysis of organic liquids and is especially valuable for gas analysis. It cannot compete with the flame ionisation detector or the electron capture detector when it comes to more sophisticated applications.

2.6.2. The Flame Ionisation Detector (FID)

The flame ionisation detector was developed several years after the katharometer, when gas chromatography was already a well established technique. Its high sensitivity and selectivity for carbon containing compounds and its wide range of linearity gave an enormous boost to the use of gas chromatography in organic analysis, and to trace analysis in particular.

There are many patterns of flame ionisation detector (for one of them see Fig. 2.6e) but they all consist essentially of a block in which hydrogen can be mixed with the effluent from a gas chromatography column and the mixed gases burned in air in a draught free enclosure. Two electrodes, maintained at a steady potential difference are placed in or near the flame and the DC current flowing between them is monitored.

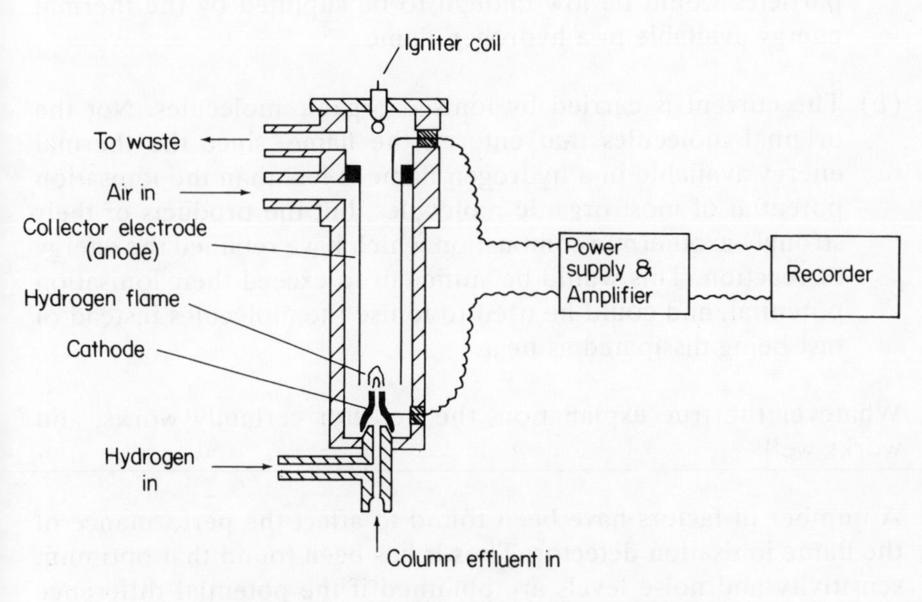

Fig. 2.6e. *A flame ionisation detector*

The current is approximately proportional to the amount of carbon in the form of volatile organic compounds which enter the flame in the column effluent, so that a graphical record of it, or chromatogram, will take the usual form of a series of peaks superimposed on a steady baseline.

The fact that the electrical conductivity of a hydrogen flame is extremely low so long as the gases entering it are pure, but increases by many orders of magnitude if the gases contain small amounts of organic material, has been explained in two ways. Which explanation you choose to believe is, at the moment, a matter of taste. The experimental evidence so far available can be interpreted as supporting either of them, and indeed, either of them will give you a good working picture of the detector by which you can interpret your results. The two contending explanations are as follows:

(*a*) The current is carried between the electrodes by charged carbon particles of colloidal dimensions which are formed in the flame by pyrolysis. The ionisation potential for such carbon particles would be low enough to be supplied by the thermal energy available in a hydrogen flame.

(*b*) The current is carried by ionised organic molecules. Not the original molecules that entered the flame, since the thermal energy available in a hydrogen flame is less than the ionisation potential of most organic molecules, but the products of their strongly exothermic combustion which have retained the energy of reaction. This would be sufficient to exceed their ionisation potential, and could be used to ionise the molecules instead of just being dissipated as heat.

Whatever the true explanation, the detector certainly works, and works well.

A number of factors have been found to affect the performance of the flame ionisation detector. Thus it has been found that optimum sensitivity and noise levels are obtained if the potential difference between the electrodes is maintained at about 200 V DC and if the jet at which the gases are burned is made the cathode, with a circular anode surrounding the flame. The temperature certainly affects the sensitivity, and ideally the detector should be situated in a separate thermostatted oven. In practice, it is often sufficient to fit the detector into the column oven, provided calibration is carried out at the temperature at which an analysis is to be performed. The relative flow rates of carrier gas, hydrogen and air affect sensitivity quite markedly if they are below a threshold. Above that threshold, which corresponds to a carrier : hydrogen : air ratio of about 1:1:10, the effect is much less noticeable and can usually be ignored. Cleanliness is important. Combustion deposits do build up on the anode and cathode and, as a result, noise levels increase. Provision is usually made for dismantling the detector so that these can be cleaned off, which is a fairly easy job. Do be careful, though, not to impregnate the electrical insulation with soap – it tends to have a somewhat deleterious effect on the insulating properties (this can also be observed if the soap from a soap bubble flow meter is allowed to drain back into the detector!).

The sensitivity of the flame ionisation detector is excellent. Routine detectors have values of around 10^7 mV mg^{-1} cm^3, although even this can be improved by careful design and operation. Within a given homologous series, the sensitivity of the detector is proportional to the percentage of carbon in the compound, but there are slight differences in the response to compounds from different homologous series. This is often within $\pm 10\%$ and correction factors have been published, although I do not think this approach is very valuable, believing that it is desirable to perform your own calibration, in order that all sources of variance can be taken into account. The sensitivity towards halogenated organic compounds is, however, significantly lower than for other compounds and depends very much on how much halogen is present in the molecule.

The linearity of the flame ionisation detector is exceptionally good, linear dynamic ranges of the order of 10^7 being quoted.

Provided that one is concerned only with organic compounds, the flame ionisation detector can be considered as a universal detector. You can be quite certain that, with one or two exceptions, if there is an organic compound present, it will be detected.

It follows from the above paragraph, that selectivity among organic compounds is poor. The exceptions to this are carbon monoxide, carbon dioxide, hydrogen cyanide, formaldehyde and formic acid, which are not detected, carbon disulphide, for which the sensitivity is very low, and carbon tetrachloride, for which it is quite low. Advantage is taken of this by using one of the liquids from this list as a solvent for samples to be analysed, since in this way the complication of a large, possibly badly tailing, solvent peak is minimised. The selectivity between organic and inorganic compounds is, however, excellent, since the latter are not detected. This is valuable, since it means that you can analyse aqueous samples for organic constituents, though of course it does mean that this excellent detector is not available for permanent gas analysis.

The flame ionisation detector is quite cheap to buy and run, although the need for three different gases is rather a nuisance, especially as, for safety reasons, they should be located outside the laboratory building. Care is needed in setting up gas flow rates and in keeping the detector clean, but in general this is not beyond the ability of most laboratory workers and the detector can be considered to be quite robust.

Π Diethylamine (bp = 56 °C), like all amines, tails very badly on most columns, giving a broad, badly shaped peak. Propanone (bp = 56 °C), like most ketones, gives a sharper, more symmetrical peak (we shall discuss the reasons for this later, in Section 3.4 of this Unit.)

You would not be very successful if you tried to determine traces of propanone in an aqueous solution of diethylamine by injecting a sample of it directly into a gas chromatograph equipped with a flame ionisation detector.

Why not?

I hope that you suggested that although the flame ionisation detector is insensitive to water, and so would not suffer interference from the water present, it is sensitive to most organic compounds, including diethylamine, and the large peak which this would produce would almost certainly interfere with the peak for the trace of propanone present.

You would need to separate the propanone from the diethylamine before gas chromatography. To do this you would probably acidify the solution, solvent extract it and then inject a sample of the extract into the gas chromatograph.

Π Which of the three following solvents would you choose to use for the solvent extraction?

(*a*) ethanol

(*b*) diethyl ether

(*c*) carbon disulphide

If your answer was:

Carbon disulphide, I think you made the best choice. It is very volatile and elutes quickly, producing a sharp peak right at the beginning of the chromatogram, so it is unlikely to overlap with peaks such as the one due to propanone, especially as the flame ionisation detector is insensitive to carbon disulphide and the peak due to it will be small.

diethyl ether, then I think that you forgot that the large amounts of ether, for which the flame ionisation detector has a normal sensitivity, would give a very large peak with a noticeable tail which would overlap with the small propanone peak and make it difficult to measure its area.

ethanol, all I can say is 'Come off it! When did your last gin and tonic form two immiscible layers?' To perform a solvent extraction the solvent chosen must be immiscible with water so that two layers are formed and can be separated.

The conflict between the value of universal and selective detector responses is at the heart of the problem which you have just been set. On the one hand, the flame ionisation detector is valuable because it responds to almost all organic compounds. On the other hand, this is a nuisance because it means that compounds of no interest, eg the solvent used to dissolve the sample, may also cause a response and so give rise to interferences. If the detector could be made to respond with greater sensitivity towards the compounds which are of interest, as it did towards the propanone compared with the carbon disulphide, it would be even more valuable.

This has been done for the flame ionisation detector by placing a solid alkali metal salt (potassium chloride or caesium bromide originally, but rubidium silicate is now more usual) just above the flame. While this does not affect the sensitivity towards simple organic compounds, it increases the sensitivity towards those containing halogens, nitrogen and phosphorus by factors of up to 500. The detector then becomes effectively selective for organic compounds containing these elements. In its most commonly used form, this *alkali flame detector* is designed to be selective for nitrogen or phosphorus containing compounds and is known as a *thermionic* or *nitrogen–phosphorus* detector (NPD). It is a valuable tool in those

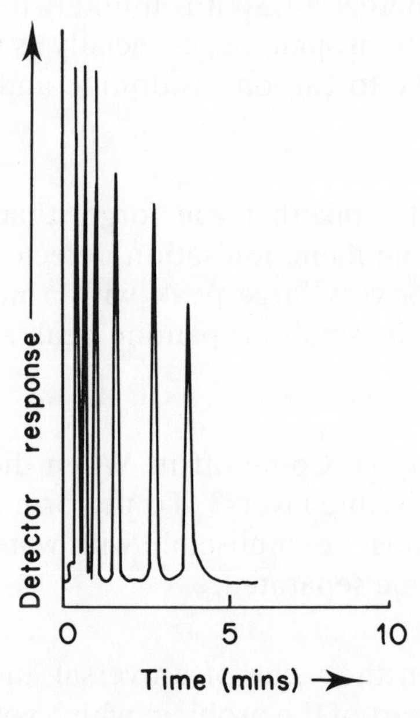

Fig. 2.6f. *FID trace*

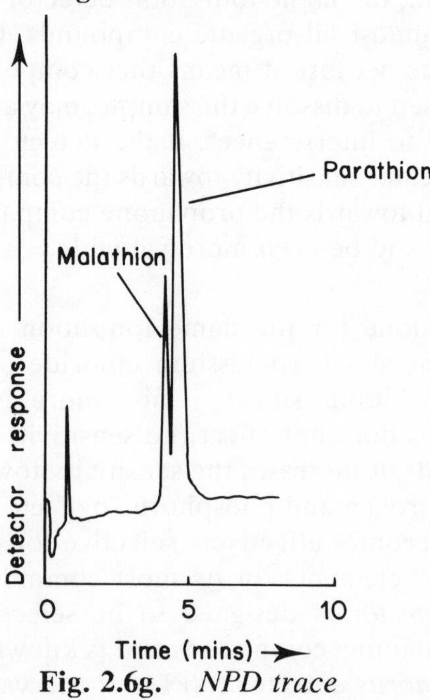

Fig. 2.6g. *NPD trace*

industries which produce and use such compounds e.g the pharmaceutical and agrochemical industries.

If you look at the two chromatograms (Figs. 2.6f and 2.6g) you will see just how powerful this detector can be under the right circumstances. Both chromatograms are of the same sample, on the same column, but the first used a conventional FID which detected all the components present and produced a very confusing pattern of peaks whilst the second used an NPD which has *ignored* the hydrocarbon solvent and other non-nitrogen containing components and produced a much simpler chromatogram. You can imagine how much it would simplify the determination of such pesticides (Parathion and Malathion).

Loosely related to these two detectors is the flame photometric detector (FPD). In this, the photoemission from the flame is monitored instead of the electrical conductivity. The flame conditions have to be carefully controlled (usually a hydrogen rich, *cool flame*) so that heteroatoms form characteristic molecular species ($S_2^{\cdot}$ for sulphur containing compounds and $HPO^{\cdot}$ for phosphorus containing compounds). The molecular emissions from these in the ultraviolet/visible region can then be selected by using a suitable filter and the intensity monitored by a photomultiplier tube. The result is a detector that is selective for compounds containing the appropriate heteroatom. It is rather more expensive than a simple FID, slightly bulkier and more difficult to set up and operate. It is less sensitive than the NPD, but its sensitivity is still high by any standards. Whilst the response for phosphorus is linear, that for sulphur is quite definitely not, being approximately proportional to the square of the sulphur concentration, which leads to some interesting calibration problems.

Like the NPD, the FPD is a useful development with very valuable applications in appropriate fields where its selectivity can facilitate an otherwise impossible analysis.

Summary

The flame ionisation detector is probably the most valuable detector for the analyst involved solely with organic chemicals to have available. It is very sensitive and very linear and ideal for trace analysis. It is quite robust and easy to use and not unduly expensive, although it does require several gases for its operation.

A variant, the nitrogen phosphorus detector, can simplify certain analyses enormously. A more distant relation, the flame photometric detector, takes this concept of selectivity still further. It is available in a form that is selective for phosphorus or sulphur containing compounds, but it can be adapted for other elements. Both of these detectors are valuable in their specialised areas, but neither is as widely used as the simple FID.

2.6.3. The Electron Capture Detector (ECD)

This detector is one of the more useful members of a large family of detectors which monitor the electrical conductivity of the effluent gas stream resulting from its exposure to the ionising radiation from a radionuclide. It has a reputation for being slightly tricky to use for routine analysis and has a far from linear response, but all this is tolerated because of its very high sensitivity and good selectivity for halogenated compounds. These properties make it especially useful in what might be loosely termed *environmental analysis*.

The detector typically consists of a small chamber within a metal block through which the column effluent is made to pass. Within the chamber are two electrodes and a β-emitting, sealed, radioactive source. There have been several geometrical arrangements of the source and electrodes within the chamber, but one (Fig. 2.6h) is reckoned to give better all round performance.

The chamber is tubular, with the source, in the form of a wide ring, around its wall. (Both ^{3}H, tritium, sealed in metal foil and ^{63}Ni have been used as a source). The tube by which the gas enters the chamber is made the anode, with the wall of the chamber acting as the cathode.

A fairly small potential difference (20 to 100 V DC) is maintained across the two electrodes and the current flowing between them

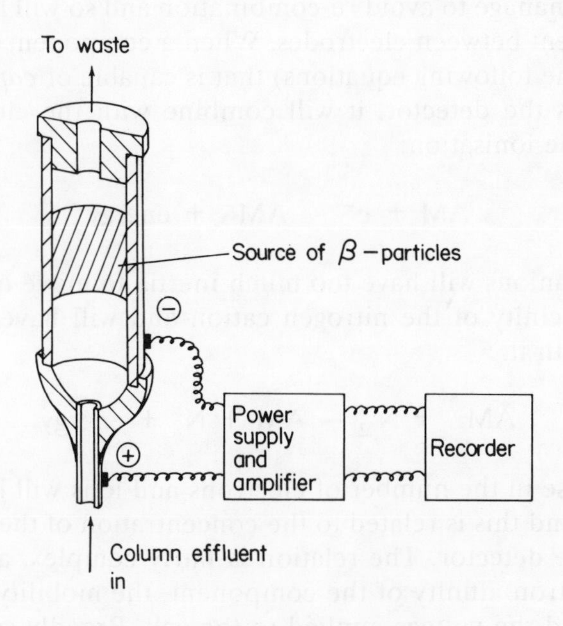

Fig. 2.6h. *Electron capture detector*

is amplified and recorded. This current is usually fairly high as a result of ionisation of the carrier gas by β-emission from the source, but it falls when a halogenated component enters the detector. The chromatogram should thus take the form of a high baseline with a trough in it corresponding to each component as it elutes from the column. It is more convenient, though, to reverse the polarity of the signal and present it in the usual form of a low baseline with peaks.

The principle of the detector is quite interesting. the carrier gas (nitrogen or argon) is ionised by the β-radiation from the source:

$$N_2 \overset{\beta}{\rightleftharpoons} N_2^+ + e^-$$

and the small, mobile electrons are attracted to, and move quickly towards, the anode before they can recombine with the nitrogen cation to form a neutral nitrogen molecule. If the potential difference between the electrodes is high enough, all the electrons produced will manage to avoid re-combination and so will help to carry a high current between electrodes. When a component (symbolised by AM in the following equations) that is capable of *capturing* electrons enters the detector, it will combine with the electrons produced by the ionisation:

$$AM + e^- \rightarrow AM^- + \text{energy}$$

The heavy anions will have too much inertia to move quickly away from the vicinity of the nitrogen cation and will have time to recombine with it:

$$AM^- + N_2^+ \rightarrow AM + N_2 + \text{energy}$$

The decrease in the number of electrons and ions will lead to a fall in current and this is related to the concentration of the component entering the detector. The relation is fairly complex, and depends on the electron affinity of the component, the mobility of the ions involved and the voltage applied to the cell. Broadly speaking, the response is proportional to concentration and is greatest for components containing very electronegative elements such as halogens.

The performance of the detector has been improved by applying the potential difference, not as a continuous voltage, but as a series of pulses of 50 V amplitude and 1 μs width at intervals of 20 to 50 μs.

Π Which of the following compounds would be readily detected by an electron capture detector?

pentane, propanone, chlorobenzene, cyclohexadiene, 1-bromobutane

If you suggested chlorobenzene and 1-bromobutane, fine. If you included propanone, you have some justification, since at the right applied potential, propanone can be detected. However, in most applications the detector would be set up to minimise the sensitivity towards oxygen-containing compounds and maximise it for chlorinated compounds so that good *selectivity* for halogenated compounds could be obtained in order to simplify the analysis. The sensitivity for hydrocarbons, which have no electronegative element to *capture* electrons, is never good.

The sensitivity of the electron capture detector is variable, depending upon the electron affinity of the compound under consideration. For compounds of high electron affinity, such as halogenated compounds for which this detector is perhaps most commonly used, it can be very high. Typical values of 10^5 mV mg^{-1} cm^3 are reported, and it is the selectivity for certain classes of compounds which makes this detector valuable.

The linearity is poor—quite the worst of any regularly used detector—though it can be improved by operating in the pulsed mode. Linear dynamic ranges of X100 are quoted for detectors using tritium sources and X50 for those using ^{63}Ni. Such poor linearity means that a very careful calibration must be carried out for each analysis and results must be read from a calibration curve which is indeed a curve. It is doubtful if this would be tolerated if it were not for the rather special selectivity and sensitivity of this detector.

That the universality is poor, and the selectivity is good, follows from the above discussion.

The detector's ease of use is more debatable. Its high sensitivity means that carrier gases must be extremely pure and dry. Both water and the next most likely impurity, oxygen, are electronegative and can cause this detector to respond and so cause a noisy baseline. Cleanliness is absolutely essential and it may be that these two factors are responsible for the electron capture detector's poor reputation for reliability. There is a slight complication in that the temperature of the tritium source should not exceed 220 °C, and whilst the ^{63}Ni source can withstand temperatures up to 400 °C, its linearity is significantly poorer.

Summary

The electron capture detector is a very specialised one. Some classes of compounds (notably halogenated compounds) it detects very well, but most it does not detect all. It is this selectivity which is the reason for its use and it is employed to simplify analyses. It is extremely sensitive but not very linear, and it is not easy to use since a lot of attention has to be paid to cleanliness and the purity of gases and solvents. It is not really one to let a novice loose on!

SAQ 2a

The sixth form at your local school has built a gas chromatograph, illustrated below. It is used to separate dichloromethane, trichloromethane and tetrachloromethane. As each of these components emerges, the flame turned from colourless to a typical *copper blue/green*.

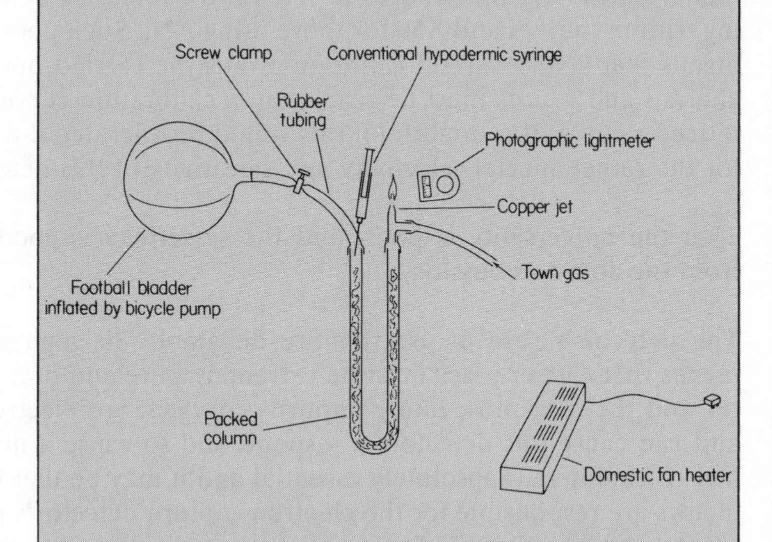

List at least six points at which this design departs from that of an *ideal* gas chromatograph.

SAQ 2a

SAQ 2b

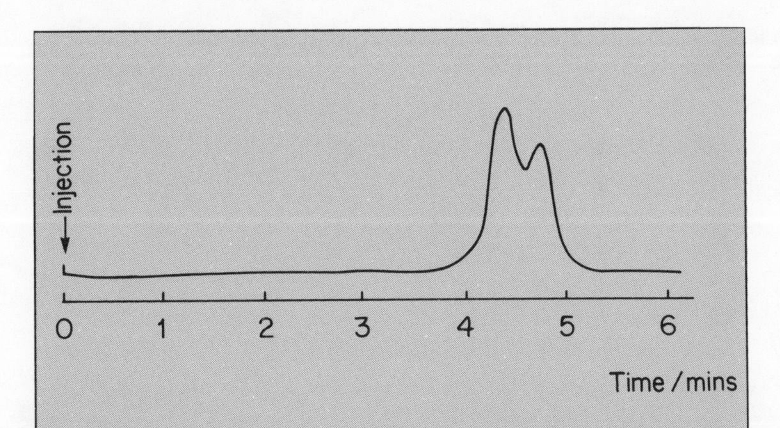

The above chromatogram was obtained with a 1 μl injection onto a 1.5 m long column at 80 °C. In order to perform a quantitative analysis for the two components, the resolution of the peaks must be improved. To do this, and perform the analysis *with the minimum delay*, would you:

⟶

SAQ 2b (cont.)

1. repeat the analysis with a smaller sample at a higher sensitivity?

2. lower the temperature and repeat the analysis?

3. repeat the analysis using a longer column of the same stationary phase at the same temperature?

4. repeat the analysis using a different stationary phase?

SAQ 2.5a

Describe a device for accurately injecting 0.05 μl of a solution of benzene in propanone into a gas chromatograph, in order to carry out a quantitive analysis.

SAQ 2.6a

Indicate the characteristics (there may be more than one in each case), chosen from the list A to D, which you associate with the detectors in the list (*i*) to (*iii*):

A. Excellent linearity

B. Good selectivity

C. Sensitivity

D. Universality

(*i*) Electron capture detector

(*ii*) Flame ionisation detector

(*iii*) Katharometer

SAQ 2.6b

Fill in the blanks in the following paragraph, choosing the missing words from the list below:

If you were asked to analyse the headspace gases in a storage tank containing gasoline, the detector you would choose would depend upon why it was being analysed. If it was only the relative proportions of the individual hydrocarbons in the vapour phase which was needed, you would use a/an............................., because it is suitably........................for hydrocarbons, but if what was needed was to know if air had been excluded by purging with nitrogen, you would use a/an.................... because it is.................

Electron capture detector

Flame ionisation detector

Katharometer

Linear

Selective

Universal

Sensitive

SAQ 2.6c

Would the following analyses be successful or unsuccessful because the named detector is inappropriate?

(*i*) the determination of traces of methylbenzene in benzene, using a katharometer.

(*ii*) the determination of carbon monoxide and carbon dioxide in furnace flue gases, to determine the efficiency of combustion, using a katharometer.

(*iii*) the determination of carbon monoxide and carbon dioxide in furnace flue gases, to determine the efficiency of combustion, using a flame ionisation detector.

(*iv*) the determination of both hydrogen cyanide and toluene diisocyanate in the products of combustion of a polyurethane foam, using a katharometer.

(*v*) the determination of both hydrogen cyanide and toluene diisocyanate in the products of combustion of a polyurethane foam, using a flame ionisation detector.

(*vi*) the determination of perchloroethylene (dry cleaning fluid) in a petroleum extract from clothing using an electron capture detector.

SAQ 2.6c

Learning Objectives

After studying the material in Part 2, you should now be able to:

- draw a labelled block diagram of a gas chromatograph;
- explain the need for flow control and flow measurement devices;
- describe commonly used flow control and measurement devices and evaluate them critically;
- identify the need for an oven and the performance characteristics needful of it;
- discuss the need for injection and detection systems and evaluate those commonly used;
- describe the column types used.

3. Columns

3.1. INTRODUCTION

All the equipment that I have described so far has only one purpose—to make it possible to use the chromatographic column. In many ways I regret the amount of time that we have spent on all the ancillary equipment (and most text books spend even more time on it). Although you may occasionally be in the fortunate position of choosing a new gas chromatograph, you more often have to use the one which is already available in your laboratory. In a larger laboratory you may have a choice of more than one, but in practice, the decision that you most often have to make is which column to use, and at what temperature, so that you get well resolved peaks which are not greatly distorted, whose retention times and areas can easily be measured. It is your ability instinctively to choose the right column and conditions that will impress your colleagues and mark you out as a good gas chromatographer.

Now we shall get down to one of the main purposes of this course—understanding columns. We can divide our discussion into two parts. Firstly, the column itself (ie the tube which contains the stationary and mobile phases) and secondly, the packing (ie the stationary phase, with or without an ancillary supporting medium which is *packed* into the column).

3.2. THE COLUMN

As we saw in the introductory chapter, two main column types are in common use—packed and capillary columns. We shall find it easier to deal with them separately.

Packed Columns

There is now quite a range of column types, though the earliest to be used, now often referred to as *conventional packed columns*, are still the most common. They are tubes between 1 m and 10 m long (in fact they are not often more than 3 m long, 1.5 m being most popular) and between 2 mm and 9 mm internal diameter (again, 4 mm id is very popular). The difficulty with very long columns is that, although they can give very good resolutions, a very high pressure has to be used to get the carrier gas to flow through them at a reasonable rate.

The earliest columns were straight. It was feared that if they were coiled, molecules travelling on the *inside* of the bend would reach the end of the column long before those on the *outside* of the bend, and this would lead to peak broadening. This effect can, in fact, be neglected if the coil diameter is at least ten times the internal diameter of the column, and most columns are now coiled so that the ovens into which they fit can be smaller and more efficient.

These columns are filled with a *packing*. This is a coarse or granular powder with particles between 0.1 mm and 0.6 mm diameter and with as narrow a size range as possible. The particle size range is commonly described not by diameters in millimeters, but by quoting the numbers of the two sieves in the US standard sieve range which at the one end is just sufficiently coarse to allow all the particles to pass through and at the other is just sufficiently fine to hold all of the particles back. Thus a packing of approximately 0.1 mm to 0.125 mm diameter is referred to as 120–140 mesh, one of approximately 0.125 to 0.15 mm diameter is 100–120 mesh and one of 0.15 to 0.25 mm is 60–100 mesh. Ideally the particles should be nearly spherical so that they pack evenly. In practice, they are nowhere near spherical.

The packing with which a column is filled can fall into one of two broad categories. It can be an uncoated solid (a simple adsorbent or one of the microporous solids which have been referred to as molecular sieves). In this case the technique of analysis which is being used is said to be *gas solid chromatography*, or gsc. Alternatively, the packing can be an inert solid coated with a thin film of a non-volatile liquid, which is the active stationary phase. The technique is then referred to as *gas liquid chromatography*, or glc.

Π Why should a column packed using a 100–120 mesh support give better resolution than an identical column packed using a 60–80 mesh support coated with the same percentage of the same stationary phase? (Reading the appropriate part of ACOL: *Chromatographic Separations*, may help if you are completely stumped).

There are three reasons, if you think about it, all deriving from the smaller particle size. Your answer should have referred to the facts that:

Smaller particles will:

1. Pack more evenly, leading to less eddy diffusion (the multiple path effect) and therefore to sharper peaks.

2. Pack more densely so that more can be packed into the column. This will give you more stationary phase in the column and the same effect as using a longer column. You would also find that the void spaces between the particles would be smaller and this would reduce the time for mass transfer between phases and hence reduce the non-equilibrium effect.

3. Have a bigger surface area per gram of solid so that the stationary phase will be spread as a thinner film. This will, again, reduce the time for mass transfer between phases and hence reduce the non-equilibrium effect.

If you had difficulty with this question, or if the rather cryptic answer outlined above does not make sense to you, then I think you ought to read the appropriate part of *Chromatographic Separations* of the ACOL series, carefully, because much of the work which we are going to cover depends on an understanding of that section.

Π Now try this one!

If a 100–120 mesh support gives better results than a 60–80 mesh one, why are 140–160 mesh supports not much used?

Did you think of the difficulty of forcing the carrier gas to flow through the densely packed column which such small particles would give, or of the pressure needed to do it?

It is not difficult to fill, or *pack* these columns and it is well worth learning how to do it. Ready packed columns are not cheap, and waiting for delivery can hold up the work of a laboratory, whereas disused columns can be emptied, cleaned and repacked fairly quickly. To pack a column, you attach a vacuum pump to the *detector* end of the column and a funnel to the *injector* end, and then slowly pour the packing into the funnel whilst vibrating the column in some way. NB: Do not forget to put a plug of glass wool (preferably silanized*) into the detector end to retain the packing!

Fig. 3.2. *Packing a column*

(For years, I have been doing this very effectively by holding an old, worn out electric razor against the column, but vibrators are

* ie—treated with a silanizing reagent to block surface polar groups which might otherwise cause adsorption of solutes.

also suitable). The vibrations and the stream of air drawn through the column by the pump help the powder to pack down tightly and evenly. When the column is completely full, you empty out the last few centimeters of packing from the injector end and put in a plug of silanized* glass wool. The column then has to be *conditioned* by removing the last traces of volatile impurities. This is done by passing a stream of nitrogen through it for 6 to 12 hours at a temperature some 20 °C hotter than its maximum working temperature. (You can actually get away with less than 6 hours conditioning in an emergency if you do not mind a bit of extra baseline noise and drift, especially if you are planning to use the column at a temperature much lower than its maximum). Your column is then ready for use.

There have been many variations of the basic packed column, each of them being designed to overcome a particular problem or to do a particular job.

For preparative gas chromatography, larger diameter columns (packed using a large particle size support) are used so that large samples can be injected onto them for separation. (One column of 300 mm id has been reported, but this is unusual. Preparative columns are usually 10 mm id).

Micro packed columns, with an id of less than 1 mm and the particle diameter of the packing reduced in proportion, and packed capillary columns (with the same id but with normal size packing particles) have been described, but they are not in common use.

Capillary Columns

The other type of column that you are likely to meet is variously referred to as a capillary or open tubular column (the latter name is the officially approved one, although the former is common usage). Such columns can be between 15 m and 100 m long and have an id of less than 1 mm (0.1 mm to 0.5 mm id is common). Unlike the columns we have discussed so far, these very narrow columns do

* ie—treated with a silanizing reagent to block surface polar groups which might otherwise cause adsorption of solutes.

not contain a packing but have their stationary phases supported on their inner walls in one of two ways. The simplest version, which was developed first, has the liquid stationary phase coated as a thin film over the smooth, inner surface of the wall of the capillary. These columns have become known as Wall Coated Open Tubular (WCOT) columns, Fig. 3.2b. The other type has the liquid stationary phase deposited onto a thin layer of small support particles (which may be porous or impenetrable) on the inner wall of the column. These are Support Coated Open Tubular (SCOT) or Porous Layer Open Tubular (PLOT) columns (the latter term also encompasses capillaries with a layer of solid adsorbent on their walls), Fig. 3.2c. The main advantage of PLOT columns is their ability to handle larger samples as a result of the larger amount of stationary phase which they contain.

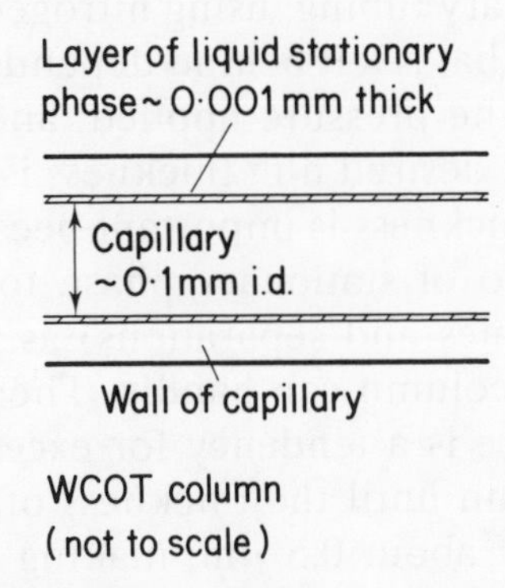

Fig. 3.2b. *WCOT column*

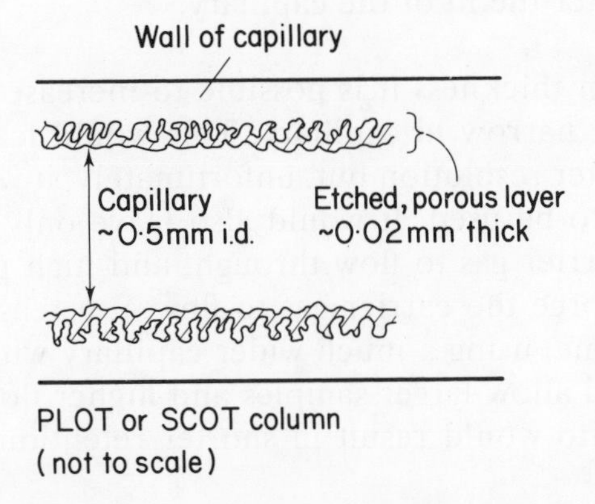

Fig. 3.2c. *PLOT or SCOT column*

The absence of packing particles in the centre of the bore of all capillary columns makes for a good, even flow of carrier gas and eliminates eddy diffusion (the multiple path effect—see ACOL: *Chromatographic Separations*), so long as the flow rate is below the threshold at which laminar flow breaks down and turbulence occurs. The resolution which can be obtained with this type of column is far better than anything a packed column can do, but of course you have to pay for it.

Capillary columns are expensive to buy, easily damaged and so difficult to make that nobody seriously contemplates doing this for themselves for routine purposes. Coating the stationary phase onto the wall of the column is difficult. This is done by blowing a solution of stationary phase in a volatile solvent through an appropriate length of clean capillary tubing, using nitrogen under pressure. The thickness of the film that is left behind depends on the concentration of the solution and the pressure applied, and getting these conditions right so that the desired film thickness is obtained is the major difficulty. The film thickness is important because it affects both the *phase ratio* (the ratio of stationary phase to mobile phase which controls retention times and separations) as well as the maximum sample size that the column can handle. There is evidence that for WCOT columns there is a tendency for excess stationary phase to bleed from the column until the thickness of the film has fallen to an optimum value of about 0.5 μm, making columns with thicker films than this unstable. Under these circumstances, the choice of film thickness is limited and the only way to control the phase ratio reliably is to alter the id of the capillary.

For a given film thickness it is possible to increase the phase ratio by using a very narrow id capillary. This would lead to longer retention and better resolution but, unfortunately, it would only allow small samples to be used. It would also leave only a small central bore for the carrier gas to flow through, and high pressures would be needed to force the carrier gas to flow at a reasonable rate. At the other extreme, using a much wider capillary with the same film thickness would allow larger samples and higher flow rates, but the lower phase ratio would result in shorter retention times and less resolution. The choice, whether to use a narrow or a wide capillary, depends very much on which criteria are more important for your particular application.

Quite recently, techniques have been developed for cross-linking liquid stationary phases in capillary columns. Such cross-linked phases are much more stable and film thicknesses up to 5 μm are quite stable. These thick films allow the use of much larger samples . They also make it possible to achieve a high phase ratio in a wider diameter column, and columns of over 0.5 mm have been introduced to take advantage of this. Their high sample capacity means that they can be used with normal micro syringes without an injection splitter—the purpose for which PLOT columns were developed.

SCOT or PLOT columns were developed in an attempt to achieve a high phase ratio in a wide bore capillary without increasing the average film thickness. The aim was to allow high flow rates and large sample sizes without the penalty of loss of resolution which would be imposed by the effect of a thick film on the non-equilibrium effect. Although many of these advantages are realised, their resolution is, unfortunately, still poorer than that of a narrow bore WCOT column. The thin layer of support particles in a PLOT column can be generated *in situ* by etching the inner surface of a glass capillary (0.25 mm to 0.5 mm id) by prolonged digestion with acid or alkali. The result is the formation of a porous layer of silica on the wall, anything between 0.005 mm and 0.05 mm thick, increasing the surface area by up to ten times its original value. The liquid stationary phase is coated onto this layer as described above. The alternative is to fill the capillary with a stable suspension of 0.01 mm diameter support particles in a solution of the stationary phase in a volatile, dense solvent. Passing a small zone furnace along the length of the capillary evaporates the solvent, depositing the support and stationary phase on the wall of the column.

Such PLOT columns can be used at flow rates of 3 to 6 $cm^3 min^{-1}$ with sample sizes of between 0.5 μl and 0.001 μl. It is just possible to employ direct injection into an injection port which passes the entire sample to the column. Such conventional injection ports usually have large dead volumes which are swept only slowly by the low carrier gas flow. There will be considerable band broadening in the injection port before the sample gets to the column. It is generally preferable to use an injection splitter, so that a higher flow rate can be passed through the injection port and a small fraction of the mixture of vaporised sample and carrier gas (3 to 6 $cm^3 min^{-1}$) passed to the column, whilst the remainder flows to waste. The danger with such a splitter is that the fraction of the mixture that enters the column will not be representative of the whole. Careful design and operation is necessary to overcome this. Such an injection splitter is essential when using WCOT columns since they require even smaller samples than PLOT columns.

This problem with injection ports highlights a more general difficulty. Capillary columns are now capable of generating such narrow, well resolved peaks that great care has to be taken over the selection of ancillary equipment to ensure that it matches them and does not limit overall performance unnecessarily. Points to be watched can range from using couplings which do not interfere with the gas flow or become contaminated, to the use of detectors and injection ports with minimal dead volumes and the selection of an integrator which responds rapidly enough and with sufficient discrimination to record the areas of such narrow peaks.

Although we shall be discussing materials used for column construction in a separate section, the development of special materials for capillary columns has made such a contribution to their performance that it would be inappropriate to leave them without discussing this subject. Early capillaries were of stainless steel or glass, although the latter was probably more popular. Glass is, however, a weak adsorbent and causes tailing, which will be especially bad when the ratio of the surface area of the support to the volume of the stationary phase is very low, as it is in a capillary column. Since the adsorptive capacity of the glass was associated with its metal silicate structure, replacing it with quartz (natural, vitrified silica) led to a great improvement. Such natural quartz had some 100 ppm of metallic impurities, and purification to reduce these to between 10 and 50 ppm led to further improvements. The ultimate performance, however, was obtained by introducing synthetic silica, with impurity levels of less than 1 ppm, and this is certainly the material of choice.

Materials

Most packed columns are made of glass or stainless steel and most capillary columns of quartz or fused silica. These materials have the advantage of being relatively chemically inert and comparatively non-adsorbent, so that they will not interfere with the chromatographic process. Like all generalisations, this is not entirely true. Hot stainless steel can catalyse the decomposition of thermally sensitive compounds which would be unaffected by glass, and biochemists in particular seem to like to use *all glass* systems, apparently for this reason. The polar surface of glass can sometimes give rise to adsorption (especially, as we have seen, in a capillary column where the surface area of the column is relatively large compared with the volume of stationary phase). This may lead to *tailing* (see ACOL: *Chromatographic Separations*, if this surprises you), and many treatments have been devised to reduce it. Silanizing is the most common treatment, but some workers have recommended applying a thin coating of a polar stationary phase to block the active sites and others have coated the glass with barium carbonate. On the whole, though, glass and stainless steel columns are reasonably trouble free. I prefer glass because you can see how well packed the column is and whether the stationary phase is discolouring or charring, but glass is much more fragile than steel and this is only a personal preference. Nylon and PTFE have both been used for making columns, but they are not all that valuable. Nylon contains a lot of absorbed water which may leak out into the gas stream at higher temperatures and anyway, nylon is permeable to water and other very polar materials. This can lead to interference under some circumstances. PTFE looks more attractive, on the face of it. It is inert and non-adsorbent. It is not much use for capillary columns though, because its non-wetting characteristics make it very difficult to coat evenly with stationary phase.

Π Aluminium tubing is cheap, has good thermal conductivity and, being malleable, can be more easily formed into a coil than glass or stainless steel. Why, then, do you think it is not much used for gas chromatography columns? (Think about the chemistry of aluminium).

I think that the main reason is the oxide film present on the surface of the aluminium. Alumina is a powerful adsorbent and a good catalyst so that the oxide film could lead to increased tailing and possibly to on-column reactions. Such reactions would result in components of the mixture, once injected onto the column, being converted into one or more products (say an alcohol changing into an alkene and water). Not surprisingly, this sort of thing can lead to a little confusion when you try to interpret your results.

Summary

There are two types of column used in gas chromatography—packed and capillary. Packed columns are the routine *work horses* of gc, being cheaper and easier to use and often giving adequate performance. Capillary columns generally give far superior resolution, and although more expensive are becoming widely used, especially for complex mixtures. Both types of column need to be made from strong materials which are non-adsorbent and chemically inert. Stainless steel and glass are the usual materials for packed columns and quartz or fused silica for capillary columns.

SAQ 3.2a

Name two materials commonly used for the tubing from which packed gas chromatography columns are made, and for each of them name two good points and one bad point.

1. ____________ Good points ____________

____

Bad point ____________

2. ____________ Good points ____________

____

Bad point ____________

SAQ 3.2b

Indicate, by circling either T for True or F for False, whether you agree with the following statements:

1. The most useful size of packed column in gas chromatography is about 1.5 m long and 4 mm id.

   T / F

2. The development of capillary columns has rendered packed columns obsolete.

   T / F

3. Since small particles of packing give so much better performance than large ones, it is better to use a packing which has a small average particle size, even if its particle size range is much larger than an alternative packing which has a somewhat large average particle size and a much smaller particle size range.

   T / F

3.3. THE STATIONARY PHASE

If, as we found earlier, the whole purpose of the rest of the gas chromatograph is to serve the column, then the stationary phase is the reason for the column's existence. All the rest, the tubing out of which it is made, the support which we shall discuss later, exist only to make it possible for the stationary phase to do its job. It is absolutely essential that you should know how the stationary phase does separate mixtures and should have a fair idea of what stationary phases are available and what they can be used for. What follows is probably the most important discussion in this Unit.

We saw, in Section 3.1, that stationary phases fall into two categories—liquids and solids, giving rise to glc and gsc respectively. Liquid stationary phases are by far the most frequently used of the two types and they are the obvious ones with which to start the discussion.

Liquid Stationary Phases

Liquid stationary phases are used in both packed and capillary columns. In either case, the properties which they need and the way in which they work are the same, and indeed, the same types of compound are used in both types of column. We can therefore discuss them without reference to the column type involved.

In order to stand any chance of doing its job, a liquid stationary phase must be non-volatile, thermally stable and chemically unreactive. If it isn't all of these things there is a good chance of materials bleeding from the column into the detector. If this happens in a variable manner, it leads to a noisy, drifting signal. Most liquid stationary phases are therefore found either to have high relative molecular masses (they are often polymers) or to be very polar molecules with strong intermolecular attractions (such as diglycerol) since such types of molecule have low volatilities. In order to achieve the necessary chemical and thermal stability they are chosen from amongst relatively unreactive classes of compounds such as saturated hydrocarbons, silicones, ethers, esters and amides. Nonetheless, for each stationary phase there will be a temperature above which the amount of material bleeding from the column into the detector will become unacceptably high. This temperature is correctly referred to as the *upper temperature limit* of that particular stationary phase. It is sometimes loosely referred to as the *temperature limit* of the stationary phase, ignoring the fact that there is also a *lower temperature limit* for most stationary phases. This is the temperature below which the liquid either solidifies or becomes very viscous. Not surprisingly, in this state it does not act as a very good solvent for the components of the mixture being separated! For most commonly used stationary phases this lower limit is well below room temperature, so it is of no great importance in normal use. Just occasionally you can get caught out as I was when I tried to improve a rather poor separation at 60 °C on Carbowax 20M by cooling the column to room temperature. My separation, far from improving, disappeared altogether. It took me a little while to work out that the stationary phase was solidifying at this lower temperature and could no longer act as a solvent for the mixture. All the components eluted together through the column without being retained at all.

Having found a range of materials which are potential stationary phases because they will both stay in the column and not deteriorate, the next problem is to choose one of them which is actually able to separate the components of the mixtures you are analysing. Separation is achieved by taking advantage of any differences both in the solubility in the stationary phase and in the volatility of the various components of the mixture.

Volatility is fairly easy to predict. It correlates quite well with the saturated vapour pressures or the boiling points of the components and has nothing to do with the stationary phase. Differences in volatility can therefore be predicted, at least approximately, by looking up the boiling points of the components to be separated. Since these are often quoted at reduced pressure, I find the nomogram shown in Fig. 3.3a very useful and keep it pinned up on my laboratory notice board.

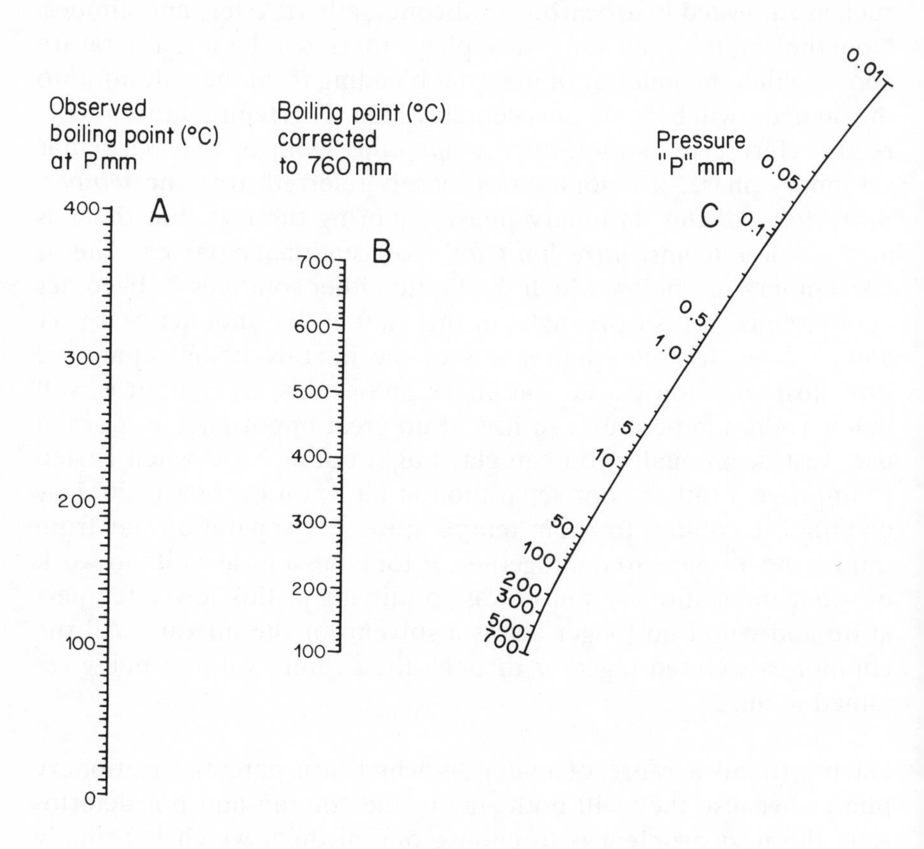

Fig. 3.3a. *Pressure–temperature alignment chart*
To estimate the boiling point at atmospheric pressure of a compound which boils at T °C at Pmm pressure, use a ruler to join the temperature T on Scale A to the pressure P on Scale C. The intersection on Scale B will be the approximate boiling point of the compound at 760 mm pressure. (e.g. dibutyl phthalate boils at 190 °C at 10 mm pressure. Joining 190 on Scale A to 10 on Scale C with a ruler intersects Scale B at 340, very close to the literature value of the boiling point of dibutyl phthalate at atmospheric pressure – 338 °C.)

Π Use the nomogram shown in Fig. 3.3a to estimate the boiling points of the following compounds at atmospheric pressure:

Ethyl 3-oxobutanoate	92 °C/40 mm Hg·
Diethyl malonate	105 °C/26 mm Hg
Ethyl 3-hydroxy-3-phenylpropanoate	129 °C/5 mm Hg
Caprolactam	141 °C/15 mm Hg
Triethyl phosphite	48 °C/13 mm Hg

Ethyl 3-oxobutanoate	185 °C
Diethyl malonate	215 °C
Ethyl 3-hydroxy-3-phenylpropanoate	285 °C
Caprolactam	265 °C
Triethyl phosphite	160 °C

I would then use these to estimate whether there was sufficient difference for there to be a reasonable chance of an easy separation and at what oven temperature I should try.

Solubility correlates well with the activity coefficient of the solute at infinite dilution, but this does not get us very much further forward. Extensive tables of activity coefficients in the liquids used as stationary phases are not readily available, so you are usually forced back onto estimating the relative solubilities of the components by looking at their structures. An organic chemist with wide experience of trying out solvents for recrystallisation and solvent extraction will have an intuitive feel for this. If, however, you are to gain any real benefit from this course, we shall need to put the appreciation of solubility onto a rational basis.

A first attempt at this might be to adopt the well established principle of *like dissolves like*. This results from the observation that saturated hydrocarbons are most soluble in other saturated hydrocarbons but not very soluble in alcohols etc., whilst alcohols are soluble in other alcohols but not very soluble in hydrocarbons. On this principle, you would predict that the solubility of the components of a mixture in a stationary phase would be in proportion to the similarity between their chemical constitution and that of the stationary phase.

Π By applying the principle of *like dissolves like*, for each of the following pairs of compounds, suggest which one will be most soluble in the stated solvent.

(*a*) naphthalene and 1-naphthol in benzene

(*b*) benzophenone and diphenylmethane in propanone

(*c*) methanol and butan-1-ol in water.

(*a*) Correct answer is naphthalene, which would be more soluble in benzene since it, too, is an aromatic hydrocarbon, whilst 1-naphthol is a phenol.

(*b*) Correct answer is benzophenone, which, like propanone, is a ketone. Diphenylmethane is a hydrocarbon.

(*c*) Correct answer is methanol. Both methanol and butan-1-ol are alcohols and share the OH functional group with water. However, they differ in the length of the hydrocarbon chain attached to it, the longer hydrocarbon chain of butan-1-ol making it more hydrophobic than methanol.

Well, if you feel that you are happy with the idea of *like dissolves like* after trying that question, have a go at the next one which applies it to a gas chromatographic situation.

Π By applying the principle of *like dissolves like* for each of the following pairs of compounds, suggest which one will elute first from a column of the stated stationary phase.

(*a*) methylbenzene (bp = 110 °C) and ethyl 2-methylpropanoate (bp = 110 °C) on squalane (a saturated hydrocarbon).

(*b*) butan-1-ol (bp 116 °C) and 4-methylpentan-2-one (bp 117 °C) on glycerol.

(*c*) hexane (bp = 68 °C) and 1-methylethyl methanoate (bp = 68 °C) on PEG-S (polyethylene glycol succinate).

You have probably noticed that for each pair the two components have the same or nearly the same boiling point, so that significant differences in volatility can be ruled out. You will have been looking at differences in solubility again. How did you get on? The correct answers were as follows:

(*a*) Ethyl 2-methylpropanoate. The stationary phase is an alkane, so that it will be a better solvent for the hydrocarbon, methylbenzene, than for the ester, ethyl 2-methylpropanoate. This will cause the methylbenzene to be retained more and so ethyl 2-methylpropanoate will be eluted first.

(*b*) 4-Methylpentan-2-one. The stationary phase is a polyhydric alcohol which will dissolve the alcohol, butan-1-ol, more than the ketone. The ketone therefore elutes more rapidly.

(*c*) Hexane. The stationary phase is a polyester, and as such will dissolve the ester, 1-methylethyl methanoate more than the hydrocarbon hexane. The hexane therefore elutes first.

Π Supposing, in the first question, I had included 2-nitroaniline and 4-nitroaniline in trichloromethane. How would you have answered that?

They are both nitroamines and so on a *like dissolves like* basis, they would appear to be likely to have similar solubilities. Yet, if you test them, you will find that 4-nitroaniline is much less soluble. Obviously, *like dissolves like* does not take us far enough.

Our organic chemist, busily choosing solvents for recrystallisation, is more likely to think in terms of the *polarity* of the solvents and the solute than to use a simple *like dissolves like* approach. He will assume that a polar solute will dissolve best in a polar solvent and vice versa. What he understands by *polarity* is a somewhat loosely defined concept which combines together the dipole moment of a compound and its hydrogen bonding ability. Thus hydrocarbons, with no dipole moment and no hydrogen bonding ability, are held to be non-polar. Water, which has a reasonably large dipole moment and very strong hydrogen bonding ability, would be reckoned to be very polar (in fact it is one of the most polar solvents we have). Alcohols, with similar dipole moments and rather less hydrogen bonding ability (only one hydrogen atom attached to oxygen), are less polar than water. Ketones and esters may have higher dipole moments than alcohols because of their carbonyl groups, but they are not strongly hydrogen bonding. They can form intermolecular hydrogen bonds only to other molecules which can supply the necessary hydrogen atom (act as hydrogen donors):

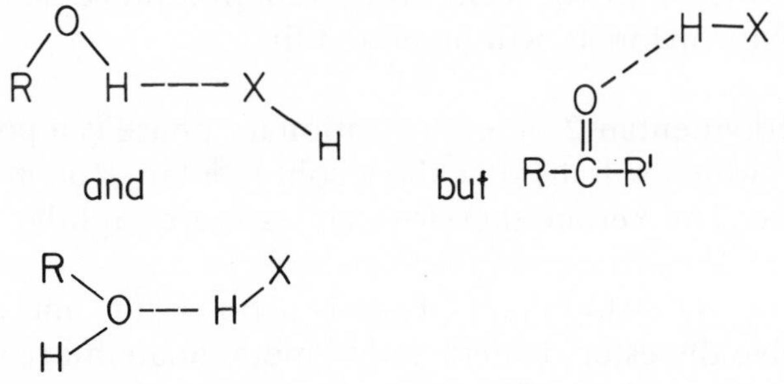

O
R H---X
H
and
R
O--- H X
H

H—X
O
R—C—R'
but

Ketones and esters are therefore less polar than alcohols. Ethers are less polar again, having a lower dipole moment than ketones because they do not have the carbonyl group and hydrogen bonding only in the same limited way, though rather less strongly because of a lower electron density on the oxygen atom:

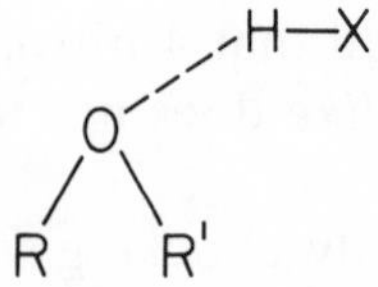

H—X
O
R R'

Halogenated hydrocarbons fit between the ethers and the hydrocarbons. They may, like dichloromethane, have a very high dipole moment, but without any significant hydrogen bonding ability they are not considered to be very polar.

Π Rearrange the following list of solvents into an order of increasing polarity:

water, iodomethane, ethyl hexanoate, octan-1-ol, hexane, ethanol, propanone.

The correct order is:

hexane < iodomethane < ethyl hexanoate < propanone < octan-1-ol < ethanol < water.

How did you get on? If you didn't get it quite right you will find it useful to read the following account of my reasoning (you might like to read it anyway).

Hexane, as the only alkane, is totally non-polar, having no dipole and no hydrogen bonding ability. Iodomethane has a dipole moment by virtue of the inductive effect of the halogen, but it still has little or no hydrogen bonding ability, and so is slightly polar. Ethyl hexanoate and propanone are both more polar because of the combination of the dipole moment and proton acceptor hydrogen bonding ability of the carbonyl groups. Ethyl hexanoate is the less polar of the two, the effect of the polar carbonyl group being *diluted* by the large, non-polar alkyl chain. Octan-1-ol and ethanol, the alcohols, are more polar because of the dipole moment and both the proton donor and the proton acceptor hydrogen bonding ability of the hydroxyl group. Again, octan-1-ol, because of its larger alkyl group, is the less polar of the two. Water, as the best hydrogen bonder of all, is the most polar.

What you have done is to re-discover an *eluotropic series*. This is a series of solvents arranged in increasing order of their ability to elute a component from an adsorbent in a chromatography column, an ability that is directly proportional to polarity. Such a series is very useful when it comes to choosing a solvent to use for elution in liquid chromatography. If your component elutes too rapidly with one solvent, you move back down the series; if it elutes too slowly, you move up the series.

If we could arrange a similar classification by polarity of the stationary phases used in glc we might find it equally useful in predicting the chromatographic behaviour of components and choosing the right column for an analysis.

Such a classification would range from the non-polar hydrocarbons through the intermediate silicone oils and greases to the slightly polar phthalate esters of long chain alcohols, then on to the more polar polyethers until it ends with the very polar polyesters. Each subgroup could then be subdivided according to small differences in polarity resulting from small differences in structure. If, say, hexane-1,6-diol were used instead of ethylene glycol in preparing a polyester stationary phase the result would be a slightly less polar material. With over 200 stationary phases listed in the catalogue of a typical specialist supplier, we obviously cannot consider here all the possible variations. Nor do we really need to. Most of the 200 or more available are variations on a few basic themes. They have slight differences in structure, leading to slight differences in polarity which might be useful if you were *fine tuning* an analysis for a particular application. Most of the time, however, you can get away with just a few basic stationary phases and compensate for any slight difference between their polarity and the polarity which would be ideal by using a longer column or a lower temperature. A stock of a basic half-dozen stationary phases will enable you to perform the vast majority of the analyses which you are likely to be asked to do. An analysis which could not be done on your basic set would necessitate ordering one of the variations, which would mean a few days delay. But then, you could hardly stock all 200 on the offchance of being asked for them, could you? You would, in fact, find that you gradually made a collection suitable for the work load of your laboratory.

Fig. 3.3b gives a selection of basic liquid stationary phases which you would find useful if you were starting a laboratory from scratch (more specialised stationary phases, eg high temperature

Stationary phase	Polarity	Upper limit (°C)	Solvent	Comment
Squalane (2,6,10,15,19, 23-hexamethyltetracosane)	NP	150	hexane	
Apiezon L (High Vacuum stopcock grease)	NP	250–300	dichloromethane	
Silicone Oils and gums eg polymethyl siloxanes				
OV 1	IP	350	trichloromethane	
DC 200	IP	200	methylbenzene	
SE-30	IP	300–350	methylbenzene	
polymethylphenyl siloxanes				
OV 17	IP	350	propanone	
SE-52	IP	300	methylbenzene	
polyfluoropropyl siloxanes				
QF1	SP	240	methylbenzene	
polycyanopropyl siloxanes				
OV 105	P	250	propanone	
Dinonyl phthalate (DNP)	SP	150	propanone	other phthalate esters also used
Polyethylene glycols eg PEG 400	P	100	methanol	RMM = 400; high proportions of terminal OH groups
Carbowax 20M	P	200	methanol	RMM = 20,000; fewer terminal group
Polyesters eg polyethylene glycol succinate (PEG-S)	VP	180	trichloromethane	
polydiethylene glycol succinate (PDEG-S)	VP	190	trichloromethane	adipates etc also used

NP = non-polar, IP = intermediate polarity; SP = slightly polar, P = polar, VP = very polar

Fig. 3.3b. *Basic liquid stationary phases*

phases, solid stationary phases, etc, to be added as work load demands).

Now let us see how we might use this concept. If we go back to mixture (*c*) hexane (bp = 68 °C) and 1-methylethyl methanoate (bp = 68 °C), (earlier in this section), we can see that we have a mixture of a non-polar compound and a moderately polar one with identical boiling points. If we chromatographed them on an intermediate or an only slightly polar stationary phase (say, silicone oil or dinonyl phthalate, respectively) they would probably not separate. Both their volatilities and their solubilities are very similar and there would be no reason for them to migrate at different rates. If, however, we used a non-polar stationary phase (say, squalane), then the non-polar hexane would dissolve readily in it and migrate slowly. The polar ester would not dissolve so well and would therefore migrate more rapidly. As a result, the mixture would be separated. On the other hand, if we used a very polar stationary phase (say, polyethylene glycol succinate) the polar ester would dissolve and be retained whilst the non-polar hexane would not dissolve so well and would migrate rapidly. Again, a separation would be observed, but the order of elution would be reversed. This pattern can be seen on the chromatograms sketched in Fig. 3.3c(i).

Things would get a little more complicated if methyl ethanoate (bp = 57 °C) were added to this mixture. To separate all three components you could use only a non-polar stationary phase. On an intermediate polarity stationary phase, the hexane and the 1-methylethyl methanoate would not be separated because of their identical boiling points. On a very polar stationary phase the hexane and the methyl ethanoate would not separate very well. The low solubility and low vapour pressure of the hexane would cause it to have a similar rate of migration to the lower boiling methyl ethanoate, the higher solubility of which would compensate for its higher vapour pressure. It is only on a non-polar stationary phase that the high solubility of the hexane would cause it to be retained more than the two more polar esters, which would both have a low solubility. The two esters would separate from each other because of the difference in their vapour pressures and would elute before the hexane. The chromatograms sketched in Fig. 3.3c(ii) show these results.

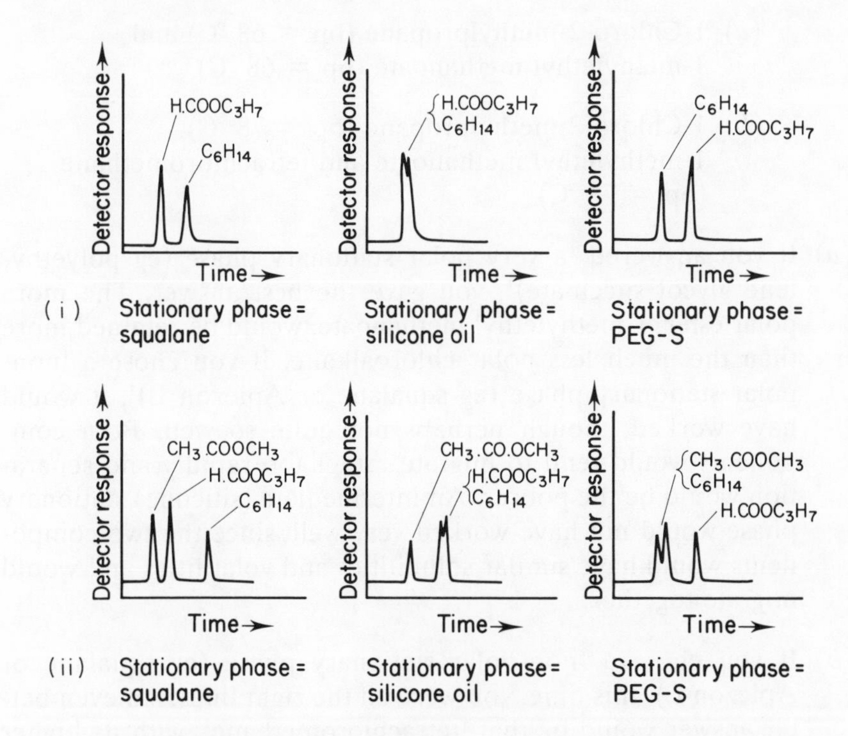

Fig. 3.3c. *Effect of stationary phase polarity on separation of:*
(i) an alkane and an ester of similar volatility;
(ii) an alkane and two esters of different volatility.

There is one complication which may have to be taken into account. As we shall see later, in Section 3.4 of this Unit, polar compounds are prone to *tailing* if they are chromatographed on columns packed with non-polar stationary phases on conventional supports. They are much less prone to *tailing* when polar stationary phases are used. If it is at all possible, it is better to use a polar column for the analysis of polar mixtures so that more symmetrical peaks may be obtained. However, tailed peaks that are separated are always more useful than symmetrical peaks that are not.

Now you try it:

Π Suggest a stationary phase capable of separating the following mixtures:

(*a*) 1-Chloro-2-methylpropane (bp· = 68 °C) and
1-methylethyl methanoate (bp = 68 °C).

(*b*) 1-Chloro-2-methylpropane (bp = 78 °C),
1-methylethyl methanoate and tetrachloromethane
(bp = 78 °C)

(*a*) If you answered 'a very polar stationary phase (eg polyethylene glycol succinate)', you gave the best answer. The more polar ester, 1-methylethyl methanoate, would be retained more than the much less polar chloroalkane. If you chose a 'non-polar stationary phase (eg squalane or Apiezon L)', it would have worked, though perhaps not quite so well. Both components would tend to migrate rather too rapidly and separation would be the poorer. An intermediate (silicone) stationary phase would not have worked very well, since the two components would have similar solubilities and volatilities and would migrate together.

(*b*) If you chose a 'non-polar stationary phase (eg squalane or Apiezon L)' this time, you were on the right lines. An even better answer would be that 'tetrachloromethane', with its higher boiling point but lower polarity than the ester 1-methylethyl methanoate, would mean that these two would migrate at about the same rate on a polar column (polyethylene glycol succinate). The lower polarity of the tetrachloroalkane would lead to lower solubility which would compensate for the lower volatility, and they would not separate. On a non-polar column the low polarity of the tetrachloroalkane would lead to it being very soluble, which together with its lower volatility would lead to a slow rate of migration. It would elute long after the other two components, which would be separated if not completely, at least reasonably well, on this phase.

Now that you have got the idea of how polarity works, it is time to take it a little further. The concept of polarity gives us a better insight into solubility and a better chance of selecting the right stationary phase for a separation. It is not unreasonable to suggest that it might be even more useful to look a little deeper into why polar solvents dissolve polar solutes best, and that is what I propose to do with you next.

Molecules associate together in the liquid or solid phase because of strong intermolecular forces of attraction between them. It is therefore energetically favourable for a solute to dissolve in a solvent only if it is possible to form attractive forces between solute and solvent molecules which are comparable in strength to those which previously existed between solute molecules and other solute molecules and between solvent molecules and other solvent molecules. If this is not possible, then solute molecules prefer to remain associated with other solute molecules with which they can form strong attractions, and solvent molecules remain associated with other solvent molecules. Dissolution does not take place; the solute is insoluble.

Since molecules with similar structures will form similar attractive forces, this is the basis of *like dissolves like*. If the solute has a similar structure to the solvent the attractive forces between their molecules will be of the same order of magnitude as the forces between two solvent molecules or two solute molecules, and dissolution is possible. It is also the basis of the way the *polarity* concept works. Molecules which have similar polarities are likely to have similar dipole moments and hydrogen bonding ability. They will therefore be attracted to one another as strongly as they are attracted to molecules of their own species by attractions between dipoles and by hydrogen bonds. Once again, they will be mutually soluble. This last idea is the key to the way forward. If we can systematically list the types of attractive forces between molecules and estimate the way in which different functional groups affect their strengths we are well on the way to predicting solubilities. It is this that we will try to do next.

The attractive forces which you will find most useful to consider in predicting chromatographic behaviour include hydrogen bonding, dipole–dipole attractions and dipole-induced dipole attractions. There are many other, often much less powerful, factors affecting the forces between solute and solvent molecules (dispersion forces, acid-base interactions, solvent/solute molar volume ratios etc) but it is the above three which you will find yourself using most of the time. A very good review of this subject has been published by S H Langer and R J Sheehan, and you might well like to read it.

Differences in hydrogen bonding probably give rise to the most dramatic differences in chromatographic behaviour. (Do you remember how hydrogen bonding seemed to be more important than dipole moments in controlling *polarity*?) Hydrogen bonds are formed between an electron deficient hydrogen atom and an electron rich atom, either in the same molecule (intramolecular) or in another molecule (intermolecular). For the hydrogen atom to be sufficiently electron deficient to form a strong hydrogen bond it needs to be attached directly to a very electronegative atom. For the electron rich atom to be sufficiently electron rich, it needs to be an electronegative atom and something of a Lewis base (able to donate a lone pair of electrons). For practical purposes, we are talking about protons bonded to oxygen or nitrogen atoms being attracted to other oxygen and nitrogen atoms. Phosphorus, sulphur and the halogens do not form significant hydrogen bonds from our point of view.

Hydrogen bonds are readily formed between hydroxyl groups:

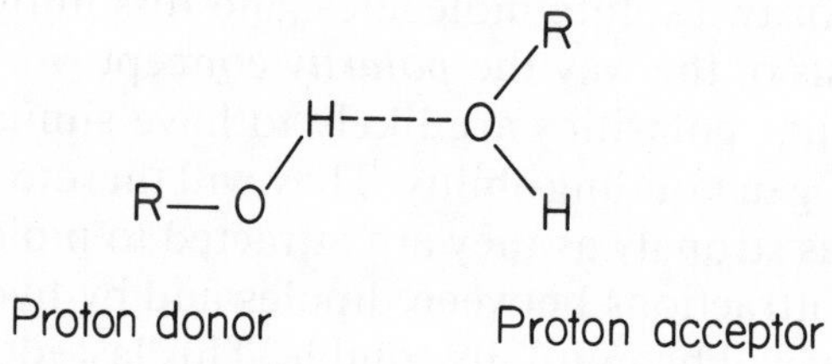

Proton donor Proton acceptor

and between amino groups:

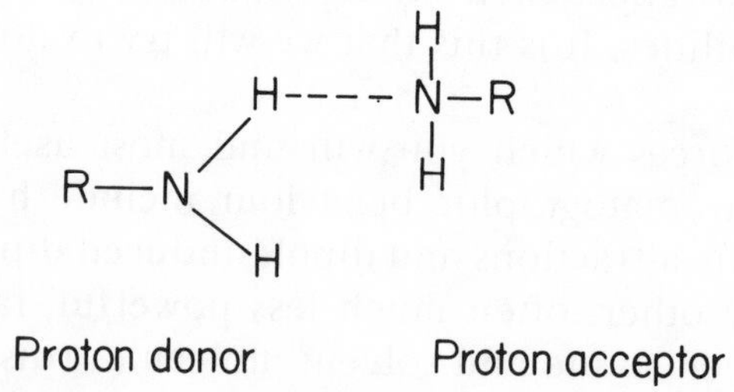

Proton donor Proton acceptor

In order to distinguish between the two molecules, one of them can be referred to as the proton donor and the other as the proton acceptor. That the hydroxyl group can form stronger hydrogen bonds than the amino group can be seen from the fact that the attractive forces in methanol cause it to liquify (bp = 68 °C) whilst methylamine remains a gas at room temperature.

Hydrogen bonds can also be formed between hydroxyl groups and the carbonyl groups of ketones, esters and amides:

H-O-R Proton donor
O
R—C—OR
Proton acceptor

H-O-R Proton donor
O
R—C—R
Proton acceptor

H-O-R Proton donor
O
R R
Proton acceptor

Hydrogen bonds cannot be formed between two ketones, for lack of the appropriate protons! Ketones, in other words, can only hydrogen bond as *proton acceptors* and not as *proton donors*. This reduction in opportunity means that they do not form such strong intermolecular attractions by hydrogen bonding as the alcohols. To judge from the relative solubilities in water of ketones and esters, though, it would seem that ketones hydrogen bond more strongly than esters. Ethers form even weaker hydrogen bonds than esters.

The common stationary phases that are important from the hydrogen bonding point of view are the polyethers (PEG, Carbowax) and the polyesters (PEG-S). The latter have high relative molecular masses and the chief functional groups are mid-chain ester groups which can hydrogen bond through their carbonyl groups. The polyethers, however, can have quite low relative molecular masses and thus a high proportion of terminal hydroxyl groups. PEG 400 approximates to the formula:

$$HO-(CH_2-CH_2-O-)_7-CH_2-CH_2-OH$$

Thus, although the polyethers are reckoned to be less polar than the polyesters, their better *proton donor* hydrogen bonding ability (due to the hydroxyl groups) can lead to them dissolving and retaining some components, whose polarity is due to hydrogen bonding, better than the polyesters.

Dipoles are most often formed in molecules when there is a displacement of the bonding electrons by an inductive effect or by resonance. The resulting centres of positive and negative charge set up an electrostatic field which can attract charges formed in a similar way in other molecules. Such attractive forces between solute and solvent molecules will encourage the solute to dissolve. Not surprisingly, larger electron displacements lead to more powerful electrostatic fields, stronger intermolecular forces of attraction and, ultimately, greater solubility.

The normal measure of the size of a dipole is the dipole moment. The dipole moments of many molecules have been measured and are available in tables, but the data are by no means extensive enough and one is usually left estimating the relative sizes of dipoles. This can be done by considering the scale of the inductive effects and/or the nature of the resonance hybrid involved. The largest dipoles result from resonance in groups involving electronegative atoms – groups such as cyano, nitro and carbonyl. Inductive effects involving singly bonded halogens, oxygen and nitrogen tend to give rise to smaller dipoles. As a rough approximation, though, the dipole correlates with the electronegativity of the heteroatom involved.

Dipole-induced dipole attractions arise when a molecule which does not have a dipole, but has polarisable electrons, is brought into the electrostatic field of a polar molecule. The field polarises the electrons, attracting them towards one end of a molecular orbital and setting up an induced dipole in the molecule. An attraction then occurs between this dipole and the original dipole, which enhances the mutual solubility of the two compounds involved.

π-Electrons are the most obvious example of easily polarisable electrons. Aromatic and unsaturated hydrocarbons (non-polar in themselves) can in fact be quite strongly retained by very polar stationary phases as a result of dipole-induced dipole attractions. This is a new concept which offers a new and valuable tool not available to us under the simple *polarity* model we were using. It should enable us to rationalise some otherwise inexplicable chromatograms. Compounds which behave in this way are called *polarisable* in order to distinguish them from *polar* molecules which have a permanent dipole.

If these three forces of attraction are not appropriate in accounting for an observed solubility, as is the case of the solubility of a hydrocarbon in another hydrocarbon, it can usually be rationalised in terms of dispersion forces. These are conceived of as the result of unspecified vibrations of the nuclei and electrons in a molecule which lead to oscillating dipoles. These, whilst cancelling each other out over the whole molecule, can induce temporary oscillating dipoles in neighbouring molecules which can vibrate in phase with them. This will happen most readily between similar molecules. The more different the molecules are, the less probable will be the effect. Dispersion forces, then, are greatest between similar molecules. Even so, they are weak compared with the other forces that we have discussed.

So, what about the 1,2- and 1,4-nitroanilines which started all this? Well, the 1,4-isomer is polar because of its ability to form intermolecular hydrogen bonds, whilst the 1,2-isomer satisfies its hydrogen bonding requirement intramolecularly:

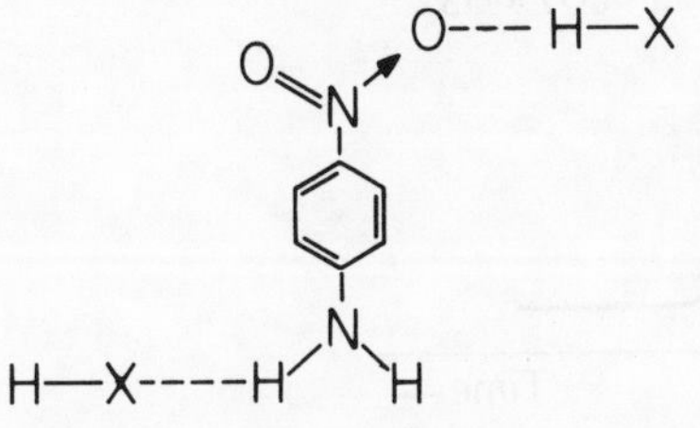

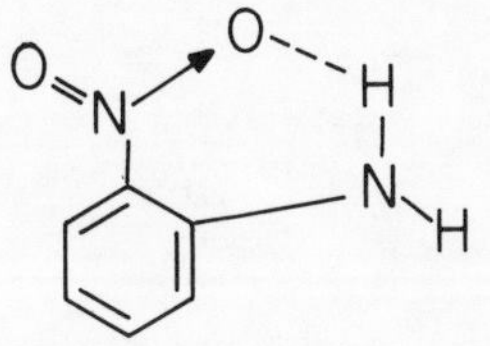

The 1,4-isomer will therefore seek to hydrogen bond to the solvent, and if it cannot do that it will hydrogen bond intermolecularly with itself and refuse to dissolve. The 1,2-isomer, because it has the stronger intramolecular hydrogen bonds, will not form intermolecular hydrogen bonds to other niteroaniline molecules and so will not need to hydrogen bond to the solvent in order to dissolve.

Now, perhaps, you should look at some examples of the use of this approach in gas chromatography.

Let us go back to mixture (*a*), methylbenzene, bp = 110 °C, and ethyl 2-methylpropanoate, bp = 110 °C. No doubt you quite rightly predicted that on a squalane (non-polar) stationary phase the methylbenzene, being less polar than ethyl 2-methylpropanoate, dissolved well and was retained more than the ester. The chromatogram would have looked like Fig. 3.3d(i).

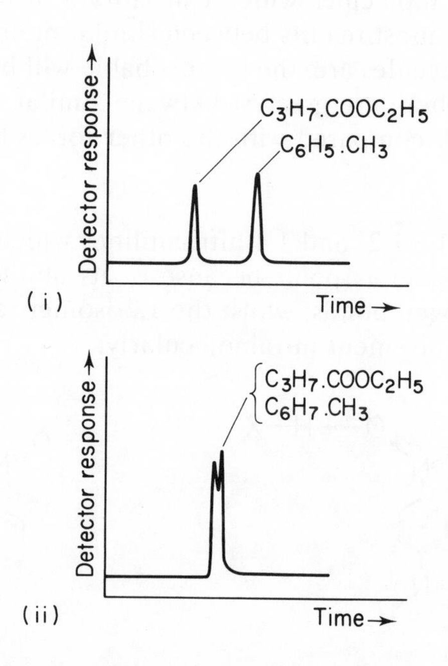

Fig. 3.3d. *Separation of an aromatic hydrocarbon and an ester of similar volatility on (i) a non-polar stationary phase (squalane), (ii) a very polar stationary phase (PEG-S)*

I was careful, at that stage, *not* to ask you what would happen on a very polar (PEG-S) stationary phase. On the information which you had at that stage, you would probably have reasoned that non-polar methylbenzene would elute long before the ester - AND YOU WOULD HAVE BEEN WRONG. In fact, the very polar polyester stationary phase PEG-S polarises the π-electrons of methylbenzene. The dipole which this induces in the methylbenzene molecule is attracted to the dipole in the polyester and this causes it to dissolve and be retained as much as the permanent dipole of the ester ethyl 2-methyl propanoate causes it to be retained. Result – practically no separation, as shown in Fig. 3.3d(ii).

Again, if you were trying to analyse an impure sample of propanone (bp = 56 °C) containing a small amount of propan-2-ol (bp = 83 °C), your first reaction might be that the difference in volatilities would be so great that you would have no difficulty in separating them. In practice, the degree of separation is not sufficient because the propanone *tails* somewhat and when you increase the sensitivity in order to be able to measure a few ppm of propan-2-ol, the tail becomes very noticeable (see Fig. 3.3e).

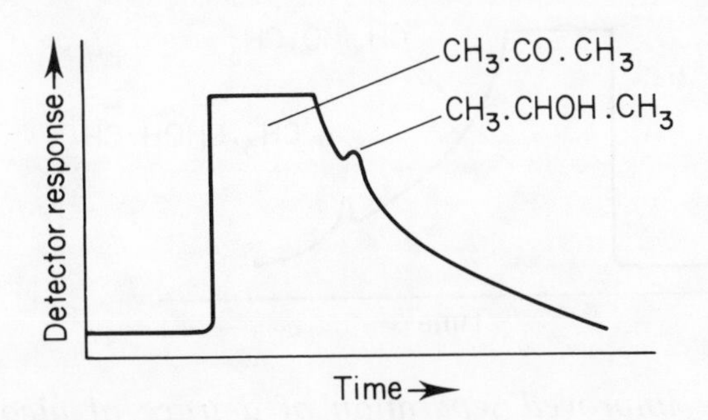

Fig. 3.3e. *Separation of a trace of an alcohol and a ketone on a non-selective stationary phase (SE-30)*

This makes it very difficult to measure the area of the propan-2-ol peak. The only solution is to change to a stationary phase which retains the propan-2-ol as much as possible whilst allowing the propanone to migrate as fast as possible.

Propan - 2 - ol (H_3C–CH(OH)–CH_3) Propanone (CH_3–C(=O)–CH_3)

A comparison of the two molecules reveals that although both molecules are polar, the polarity of propan-2-ol will be, in large measure, due to hydrogen bonding whilst that of propanone will be mainly due to dipole–dipole attractions. Consequently, you would choose a stationary phase with a strong hydrogen bonding capability and only a moderate dipole. This would differentiate between the two components as much as possible, pulling the propan-2-ol back off the tail of the propanone. A low molecular weight polyether (PEG 400) is the obvious one to choose, since this has a high proportion of terminal hydroxyl groups for hydrogen bonding and only a modest dipole due to the ether group. The result would look like Fig. 3.3f.

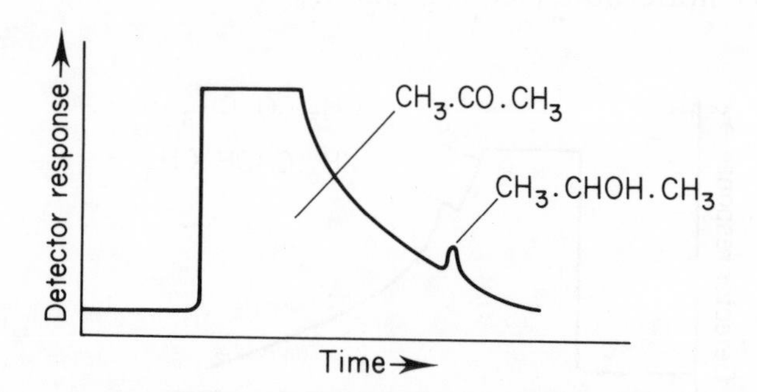

Fig. 3.3f. *The improved separation of a trace of alcohol and a ketone on a polar stationary phase capable of hydrogen bonding (PEG 400)*

Now you have a go!

Π Against each solute/stationary phase pair in List A, circle the letter(s) corresponding to the attractive force(s) from List B (there may be none or there may be more than one) that are likely to be significant in causing solubility.

List A

1. methanol/squalane — a b c d

2. methanol/PEG 400 — a b c d

3. chlorobenzene/dinonyl phthalate — a b c d

List B

a. hydrogen bonding

b. dipole/dipole attraction

c. dipole/induced dipole attraction

d. dispersion forces

1. methanol/squalane. a b c d

That's right! None of them: methanol is virtually insoluble in squalane and would have an extremely short retention time. As an alkane, squalane has no groups capable of hydrogen bonding, has no dipole and no polarisable groups, and it is so different in structure that dispersion forces will be exceedingly weak.

2. methanol/PEG 400 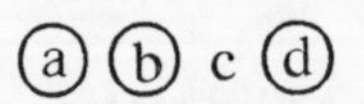

Both of them have hydroxyl groups, so hydrogen bonding and dipole/dipole attractions will be strong, and with such similar groups dispersion forces will be not inconsiderable. Neither molecule has a polarisable group, so that induced dipoles are not going to be observed, but with the other three forces being so strong, methanol should be quite soluble in PEG 400. (Now have a look at Fig. 3.3b and see what solvent is used to dissolve PEG 400 when it is being coated onto a support).

3. chlorobenzene/dinonyl phthalate 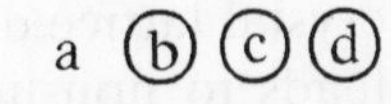

Well, hydrogen bonding is out, but both chlorobenzene and DNP have small dipoles and both have a polarisable aryl group, so dipole and induced dipole effects should occur for both of them. On top of that, dispersion forces should be quite large because of the similarity (both have a slightly polarised aryl group in their structures). The retention time of chlorobenzene should be quite long on this phase.

Solid Stationary Phases

In general, these have been much less popular than liquid stationary phases because they frequently lead to long retention times, badly tailed peaks and poor reproducibility. Nontheless, I find them quite interesting because they show selectivity amongst the components of a mixture which is far in excess of anything of which liquid stationary phases are capable. They have been very useful to me on occasions, but whether you can take advantage of their potential depends very much on whether you can tolerate or overcome the disadvantages listed above. By and large, you are most likely to use them for separating gases – something which cannot easily be done with liquid stationary phases. This is a somewhat more specialised area of gas chromatography than the analysis of organic liquids, but it is still an important application.

Solid stationary phases can be loosely divided into two groups: those which operate by adsorption and those which operate on the *molecular sieving* principle. Naturally it is not as clear cut as that, and some molecular sieves also act by adsorption, but it is convenient to treat the two separately.

Adsorption onto a solid surface is not a uniform process. Some areas of the surface adsorb molecules more readily and more strongly than others and it is in these areas that the adsorbed molecules cluster most thickly. Such areas are known as *active sites* and they tend to be found at physical irregularities (cracks and crevices etc) and at crystal lattice defects on the surface. It is this non-uniformity which leads to non-linear adsorption isotherms and tailing, as explained in the ACOL Unit: *Chromatographic Separations*. If you feel that you need to re-familiarise yourself with this, you should read the relevant section in that Unit before proceeding

The forces responsible for adsorption are remarkably similar to those responsible for solubility – hydrogen bonding, dipole–dipole attraction, dipole-induced dipole attraction, dispersion forces, etc. In the context of gas chromatography you may find that acid/base interactions (the donation of a lone pair from an atom of one molecule to an electron deficient atom in another) assumes a greater importance, but otherwise they are very much the same. The extra selectivity of adsorption, compared with solution, probably arises from the rigid geometry of the solid surface. The atoms of the adsorbent which are responsible for attracting the adsorbed molecule are held rigidly in place by their crystal lattice. Their ability to attract the relevant atoms of the adsorbed molecule will depend upon the geometry of that molecule much more than in the case of attractions to the flexible, movable liquid molecules.

Molecular sieves, at least to some extent, function because of the many small molecular sized pores which penetrate their structures. Small molecules are able to penetrate into these pores and become adsorbed on their inner surfaces. Larger molecules are less able to penetrate becoming adsorbed on the outer surfaces, and so are eluted more quickly. The order of elution from columns of such stationary phases is roughly the order of decreasing molecular size or relative molecular mass. Of course it is never that simple. Molecular sieves are also adsorbents and differences in ease of adsorption of the molecules will be superimposed on this pattern. Still, you can manage quite a number of otherwise impossible separations with molecular sieves.

Common solid stationary phases are:

Alumina

Alumina (Al_2O_3) is a powerful adsorbent. It can hydrogen bond through hydroxyl groups formed on its surface by hydration, attract by dipole–dipole and dipole-induced dipole attractions through the same groups and through oxygen atoms in the surface, and electron deficient aluminium atoms can accept electron donation (acid/base interaction). It does an excellent job of separating the lower alkanes and alkenes and of analysing mixtures of freons. However, components are very susceptible to tailing and although this can be partly overcome by depositing inorganic salts onto its surface, you can never get truly symmetrical peaks.

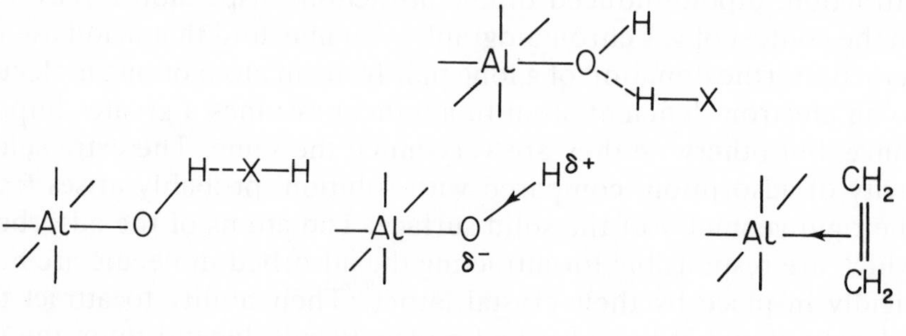

Carbon Black

Carbon black has long been used as an adsorbent for gsc. In principle it should have no polar groups in its surface so that adsorption would be due solely to dispersion forces. The possession by the components of the mixture of functional groups, π-bonds or lone pairs of electrons, is therefore irrelevant to their adsorption. This should be controlled by the size, shape and polarisability of their molecules, and so, although selectivity between molecules could still be very high, it should be much less species dependent.

In practice, simple carbon black does possess a variable number of polar groups on its surface and these lead to badly shaped peaks and unreliable performance. Not surprisingly, carbon black in this form was used only sporadically. If, however, carbon black is heated in an inert atmosphere to about 3000 °C many impurities are removed and the surface undergoes a transformation into graphite. Such graphitised carbon black (GCB) is a much improved adsorbent. Even so, there are still physical irregularities and a few small areas of polar groups in the surface which have the effect of deforming peak shapes. This can normally be largely overcome by coating the adsorbent with a small amount (up to 1 or 2%) of a liquid stationary phase (often PEG). Whether you should think of this as just blocking the active sites or whether you should now think of the mechanism as a combination of partition and adsorption is arguable. What is significant is that vastly improved peak shapes and somewhat reduced retention times result and the stationary phase can be used to separate a wide range of organic compounds, both gases and liquids. These include saturated hydrocarbons, amines, phenols and aromatic acids.

Zeolites

These are the original alumino-silicate molecular sieves. They are also powerful adsorbents, adsorbing water, carbon dioxide, etc, irreversibly below temperatures of around 200 °C. (This means that columns must be protected from the atmosphere when not in use). Of course, they cannot be used for the gas chromatography of these gases. They are, however, excellent for separating the noble (inert) gases, oxygen, nitrogen, carbon monoxide and other gases and for separating straight and branched chain hydrocarbons, all on the basis of size.

They are marketed by Union Carbide as the Linde range of molecular sieves.

Silica Gel

Silica gel is slightly odd in that, although it is porous, and its pore size certainly influences its performance as a stationary phase, it operates fundamentally as an adsorbent not as a molecular sieve. Hydroxyl groups on the surface seem to be the main sites of adsorption; dispersion forces, dipole and induced dipole interactions and hydrogen bonding are all important, but there is some evidence that, in hydrogen bonding, it acts much more effectively as a proton acceptor than as a proton donor.

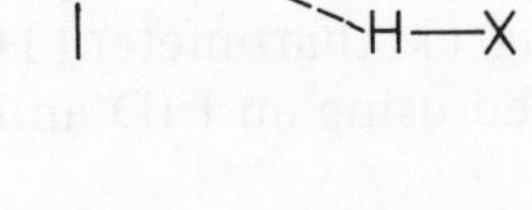

$$-\mathrm{Si}-\mathrm{O}(\mathrm{H})\cdots\mathrm{H}-\mathrm{X}$$

It finds many uses in gas analysis, including the separation of carbon dioxide from other gases.

Porous Polymers

Styrene can be polymerised under conditions which lead to beads of porous, crosslinked polystyrene. Again, pore size affects chromatographic performance, but the mode of action is adsorption rather than molecular sieving. Adsorption takes place on the surface of the polymeric aromatic hydrocarbon within the pores, rather as it does on the surface of graphitised carbon black. The pores seem chiefly to ensure a very large surface area for adsorption to take place on. If anything, the polystyrene surface has even fewer extraneous polar groups than GCB, and peaks are very symmetrical. You can get excellent chromatograms of such components as water, ammonia, the lower alcohols and amines, all with very little sign of tailing.

They are marketed by Waters Associates under the name Porapak and by Johns–Manville as Chromosorbs.

Tenax-GC is another porous polymer which is available in bead form for packing gas chromatography columns. It is based upon 2,6-diphenyl-p-phenylene oxide and has a greater thermal stability than polystyrene. Slight column bleeding is noticed above 320 °C, but this does not become excessive until a temperature of 375 °C is reached. Like polystyrene, it gives very symmetrical peaks and with its high temperature limit it is excellent for separating higher boiling polar compounds such as aromatic amines, phenols, glycols, etc.

Perhaps the best way to get some idea of the separations that are possible by gsc is to look at the following chromatograms. They *say* much more than pages of writing. They were all obtained, using a routine instrument fitted with conventional, packed columns (1.5 m × 4 mm id) and a carrier gas flow rate of 45 cm^3 min^{-1}. Figs. 3.3g, h and i were obtained using a katharometer (TCD) and helium carrier gas; Fig. 3.3j was obtained using an FID and nitrogen carrier gas.

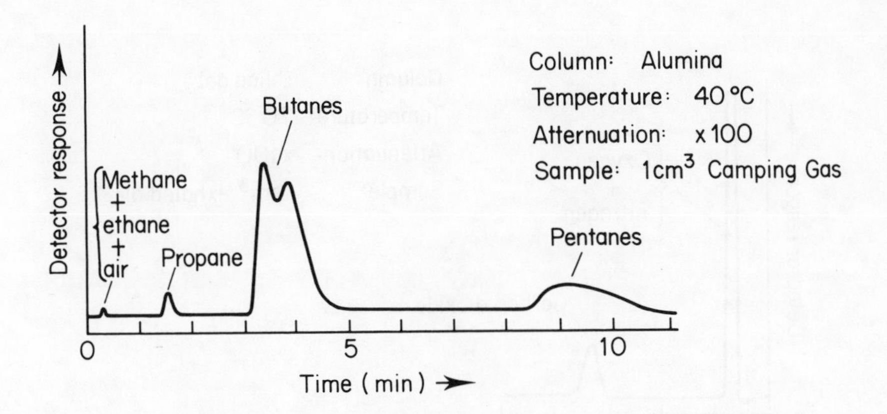

Fig. 3.3g. *Chromatogram of camping gas*

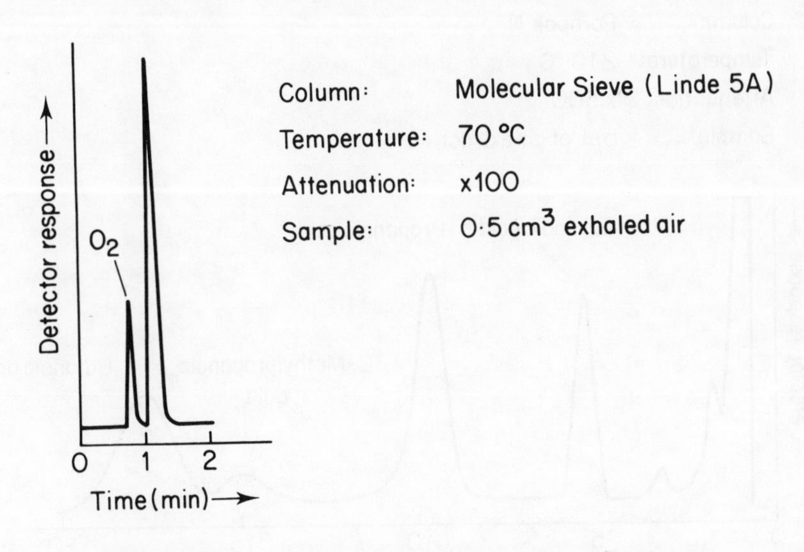

Fig. 3.3h. *Chromatogram of exhaled air – I*

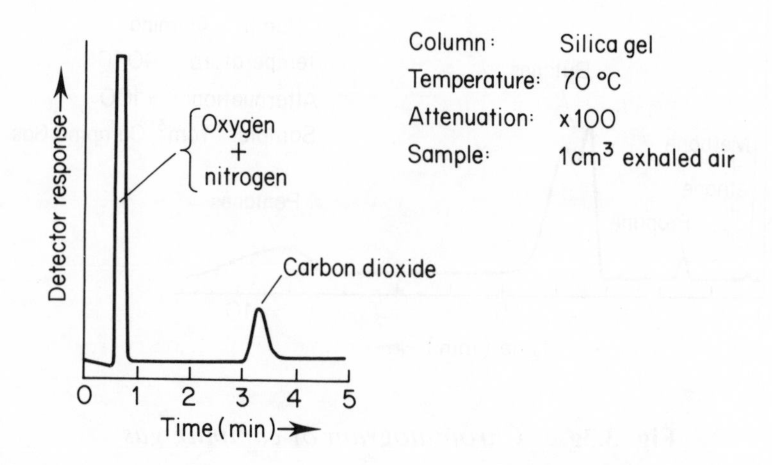

Fig. 3.3i. *Chromatogram of exhaled air – II*

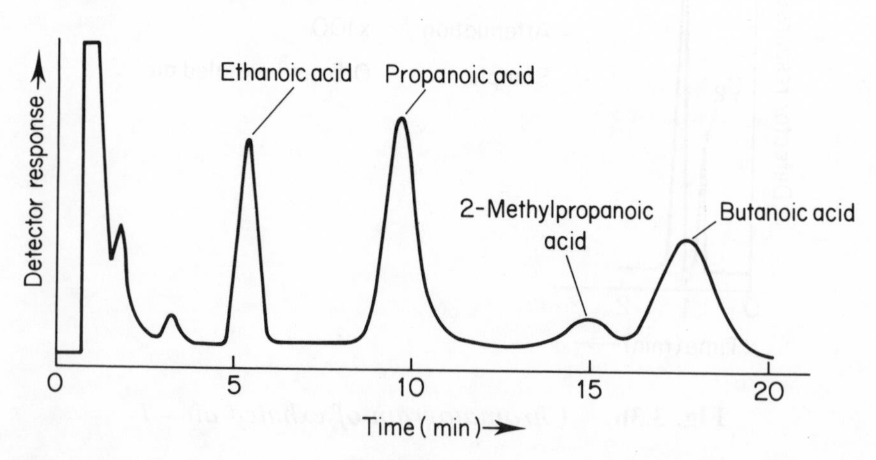

Fig. 3.3j. *Chromatogram of the alkanoic acids extracted from raw sewage sludge*

∏ Would you expect benzene (bp = 80 °C) to elute before, at the same time or after cyclohexane (bp = 80 °C) on an alumina column?

If you answered 'after', well done. If you thought it would elute before or at the same time, then you had probably forgotten that the very polar groups in the surface of alumina would induce a dipole in benzene by polarising its π-electron cloud. This would cause it to be more strongly adsorbed than cyclohexane, which has only much less polarisable σ-bonds.

∏ Which of the solid stationary phases that we have discussed would you choose to separate the following mixtures?

1. Methane, ethane, propane, butane.
2. Helium, neon, argon.

1. If you chose a graphitised carbon black, preferably modified with a little PEG 400, you have made the best choice. It will give more symmetrical peaks than alumina, which is the other phase capable of separating them.

2. You should have have chosen a Linde molecular sieve. The noble gases offer no interactions with other molecules which you can use to differentiate between them – you can exploit only their size difference.

Summary

Liquid stationary phases are in much more common use than solid stationary phases, probably because of their great reliability. Liquids that are to be used as stationary phases must be non-volatile and chemically and thermally stable. They are best classified in terms of their polarity, that is as *non-polar, intermediate, moderately polar* and *very polar*. Their ability to dissolve and retain components correlates quite well with the similarity between the polarity of the compound being chromatographed and the stationary phase. For finer tuning and a more sophisticated choice of stationary phase it is necessary to look more deeply at the reasons underlying solubility and to consider the main attractions between solvent and solute molecules. These are hydrogen bonding, dipole–dipole attractions and dipole-induced dipole attractions, with dispersion forces making a lesser contribution.

Solid stationary phases, whilst being less commonly used, are attractive because they offer much greater selectivity and in fact, are often essential for the analysis of gases. They are, however, easily contaminated and retention times and separations may not be very reproducible. Selectivity will be due to differences in hydrogen bonding ability, dipole and induced dipole interactions and to dispersion forces or to differences in molecular size.

SAQ 3.3a

Place the following stationary phases in order of increasing polarity:

polyethylene glycol succinate (PEG-S),

polyethylene glycol – relative molecular mass 400 (PEG 400),

polyethylene glycol – relative molecular mass 20,000 (PEG 20M),

hexadecane,

tritolyl phosphate,

polypropylene glycol adipate (PPGA).

SAQ 3.3b

Benzene, cyclohexane and ethanol are to be separated by glc. Given that they all boil between 70 °C and 80 °C, indicate by circling either T for True or F for false whether you agree with either of the following statements.

1. On a squalane stationary phase at 70 °C the order of elution would be:

   ethanol, followed by benzene, followed by cyclohexane.

   T / F

2. On a polyethylene glycol succinate (PEG-S) stationary phase at 70 °C, ethanol would elute after both benzene and cyclohexane.

   T / F

You might like to consider what would happen if PEG 400 were used as a stationary phase instead.

SAQ 3.3c

For the three mixtures below, select in each case the answer (*a*), (*b*) or (*c*) which you think is the correct order in which the named components of the mixture will elute on the given stationary phase.

1. Cyclohexane (bp = 81 °C) and cyclohexene (bp = 83 °C) on dinonyl phthalate (DNP).

   (*a*) More or less together
   (*b*) Cyclohexane then cyclohexene.
   (*c*) Cyclohexene then cyclohexane.

2. Methoxybenzene (anisole, bp = 154 °C) and 1-methylethylbenzene (cumene, bp = 152 °C) on polyethylene glycol (PEG 400).

   (*a*) More or less together.
   (*b*) Anisole then cumene.
   (*c*) Cumene then anisole.

3. Hexane (bp = 68 °C), 1-methyl-1-(1-methylethoxy) ethane (diisopropyl ether) (bp = 68 °C) and propan-2-ol (bp = 83 °C), on polyethylene glycol succinate (PEG-S)

   (*a*) Hexane, diisopropyl ether, then propan-2-ol.
   (*b*) Diisopropyl ether, hexane, then propan-2-ol.
   (*c*) Hexane, propan-2-ol, then diisopropyl ether.

SAQ 3.3d

Explain, in the space below, why gases are more likely to be analysed by gsc than glc (and if you just say 'because gsc separates gases better than glc' I shall ask you why it does, so you might as well get down to fundamental reasons to begin with!).

SAQ 3.3e

If you were asked to determine the concentration of carbon monoxide and carbon dioxide in a boiler flue, which of the following stationary phases would you use?

1. To determine carbon monoxide __________

2. To determine carbon dioxide __________

Stationary phases:

Alumina, silica gel, graphitised carbon black, Linde molecular sieve, porous polystyrene (Porapak).

3.4. THE SUPPORT

Since the only function of the support is to immobilise the stationary phase, it is not difficult to suggest a number of properties that it should have:

1. Chemical inertness, thermal stability, non-adsorbency.

2. Mechanical strength.

3. Uniform particle size (spherical would be ideal).

4. Large surface area per unit volume.

The supports in common use meet these criteria to varying degrees.

By far the most important group of supports are the diatomaceous earth based materials (diatromite, kieselguhr, celite, chromosorb, Embacel, Sil-O-Cell, crushed firebrick). Diatomaceous earth is a mineral deposit derived from the siliceous, fossilised remains of diatoms (small sea creatures whose descendants are still to be found in plankton). Under a microscope, the structure of the creatures from which it has been formed can be seen. The particles can be seen to be far from spherical, with quite a wide range of sizes and shapes. Diatomaceous earth is very porous and therefore has a high surface area, enabling it to absorb a large amount of liquid stationary phase (up to 20–30%) whilst still remaining a free flowing powder. Chemically, diatomaceous earth consists of about 95% SiO_2, the remainder being Al_2O_3 and other metal oxides. Whilst it is chemically inert, it is a weak adsorbent, and so it can interfere with glc. It does not naturally have great mechanical strength, and so it is often modified to improve this property. This is usually done by heating at 900 °C, either with a little sodium carbonate (giving a 'white' support) or with a little clay (giving a 'pink' support). The latter are unpopular and almost never used now, because although they are mechanically stronger and more porous, they have a much higher residual adsorptive capacity, which leads to badly tailed peaks. Even the white supports have some residual adsorptive capacity, and they are usually treated to reduce it. This treatment normally takes the form of heating the support with concentrated hydrochloric acid for 30 minutes, washing it with water until the washings are neutral, then washing it with methanol and finally drying it. This seems to remove many of the trace metal oxides as well as much of the fine dust and many sharp edges. The support is then often 'silanized' by treatment with one of the several reactive alkyl silane derivatives (see Section 5.3). In this way, hydroxyl groups on the surface of the support, which are sites for adsorption, are converted into silyl ethers, which are not, eg:

$$-Si-OH \xrightarrow{CH_3.CO.N(SiMe_3)_2} -Si-O-SiMe_3$$

Such supports (Acid Washed, Silanized) are certainly not perfect, but they are probably the most satisfactory supports that we have. It has been said that 'all problems can be solved on kieselguhr; there is no real need for any other support' (how about that for an essay title for your tutor to give you –'Justify the above statement'?).

Other supports certainly have been, and are, used. Porous PTFE, porous polystyrene and glass beads have all been tried in attempts to find non-adsorbent supports. The organic polymers are expensive and tend to soften and clog up the column when coated with stationary phase, even at quite moderate temperatures. Glass beads do not do this, but the stationary phase does not coat them evenly, being drawn to the points of contact by its surface tension, where it forms quite thick fillets (Fig. 3.4).

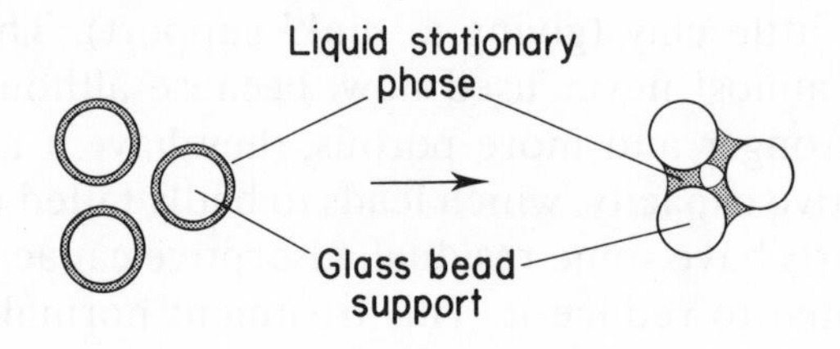

Fig. 3.4. *Uneven coating of glass beads*

This leads to slow mass transfer (see ACOL: *Chromatographic Separations*), non-equilibrium and broadening of peaks. In practice, this limits glass beads to a stationary phase loading of about 3%. Even at 3% loading, the beads are a little sticky and difficult to pack into a column. It is best to coat them *in situ* by a technique similar to that used for capillary columns (blowing a solution of the stationary phase through the column with nitrogen under pressure).

Coating the Support

This is not difficult, but you need to take a little care when doing it. You can buy supports already coated with most stationary phases, but it is worth learning to coat your own because you then have much more flexibility over the per cent loading and the use of mixtures of stationary phases.

To coat a support, you first weigh out, separately, appropriate weights of support and stationary phase. The stationary phase is then dissolved in a volatile solvent (most suppliers of stationary phases list the solvent appropriate to each of them. See Fig. 3.3b). It is essential that the solvent you use should be pure, because the next step is to add the support to this solution, making a wet slurry, and then to evaporate the solvent. Because of the large quantity of solvent used, even traces of non-volatile impurities can result in quite large amounts of contamination being left behind in the stationary phase, where they may well affect its performance. To achieve as homogeneous a packing as possible, the evaporation needs to be carried out with gentle stirring and heating. I usually do this using a rotary evaporator, but I always set it to the slowest possible speed of rotation to stop it acting as a ball mill and grinding the support to an even finer powder. In any case, I usually re-sieve the powder at this stage to remove any *fines* or *conglomerates* and then finish drying it in an open dish in a warm oven.

There is a little confusion over the weights which you should take to get a 10% loading. Some authorities (and I agree with them) recommend 90 g of support plus 10 g of stationary phase. Others (and this does not seem very logical) have recommended 100 g of support and 10 g of stationary phase. This latter seems to me to be a 9.1% loading. The difference would probably not prevent you getting a separation, but it would affect all your retention times.

10% is a common loading to use. It is a compromise between, say, 20% and 5%. The former gives a larger capacity, allowing larger samples to be injected, but it also gives rather long retention times and, because the film is thicker, slow mass transfer and severe peak broadening. It also represents the effective upper limit for diatomaceous earth supports, since at loadings much above 20% the packing starts to become rather sticky. A 5% loading certainly gives sharper peaks because of the faster mass transfer, but it also gives shorter retention times and smaller differences in retention times, so that a longer column could well be needed to get the same resolution.

Lower loadings than 5% are used occasionally (as little as 1%). They might be used if you had a very high boiling mixture to analyse and did not want the retention times to be excessive. This is quite useful for such samples as steroids and lubricating oils. Just occasionally, with so little stationary phase present to act as a solvent, adsorption by the support becomes more noticeable and you may get rather surprising retention times and badly tailed peaks. You are most likely to observe this sort of effect if you are using a non-polar stationary phase. Polar stationary phase molecules will be attracted to the active (polar) sites of adsorption on the support. Once adsorbed, they will block the active sites, preventing adsorption of analyte molecules so that the support does not interfere with the chromatography and symmetrical peaks result. With a non-polar stationary phase, this mechanism is not available and so tailed peaks and other results of adsorption are more likely to be observed.

Summary

The support onto which the stationary phase is coated in packed columns needs to be mechanically strong, chemically inert and non-adsorbent. The particles should be small, of a narrow particle size range, regular shape and ideally should be porous so that they have a large surface area. Modified diatomaceous earth supports probably come closest to the ideal, especially if they are washed with acid and silanized to reduce adsorption. The stationary phase (usually 10% of the combined weight of stationary phase and support) is deposited onto the support from solution by evaporation of the solvent.

SAQ 3.4a

List three respects in which diatomaceous earth falls short of the ideal as a support for gas chromatography.

1.

2.

3.

SAQ 3.4b

On which of the following columns is the *tailing* of a phenol peak likely to be worst?

1. 10% Apiezon L on diatomaceous earth.

2. 2% Apiezon L on diatomaceous earth.

3. 10% PEG 400 on diatomaceous earth.

4. 3% PEG 400 on PTFE beads.

Learning Objectives

After studying the material in Part 3, you should now be able to:

- describe packed and capillary columns;

- describe the preparation of packed columns;

- discuss the mechanism by which common liquid and solid stationary phases retain the components of a mixture;

- understand the role of the support in gas chromatography.

4. Choosing the Other Parameters

4.1. INTRODUCTION

We have spent quite a long time talking about how stationary phases *work* and how you go about choosing the right one for an analysis. This is as it should be. Getting the right stationary phase is the first, and most important, decision that has to be made. The other decisions will lead to an improvement in the basic separation, but they cannot create a separation where, because of the properties of the stationary phase, one does not already exist.

What, therefore, are these decisions? Foremost, I would say, is the choice of temperature for the column oven. After that comes sample size and detector attenuation (which are of course inter-related), and then flow rate and injection heater temperature. Somewhere amongst these comes the length of the column, which may be absolutely vital in a few analyses, eg where a long column is needed to achieve resolution or a shorter column is needed to avoid an excessively long analysis time. However, it is unrealistic to expect to stock a wide range of column lengths in all stationary phases. The decision to order a new, shorter column in order to save analysis time would be taken only if the analysis was going to be repeated frequently.

My own approach is to start with a fairly standard set of conditions:

Stationary phase:	as chosen; 10% on 100–120 mesh diatomaceous earth (AWDS).
Column:	1.5 m long; 4 mm id.
Temperature:	about 20 °C below the average boiling point of a mixture, or, if only a single component in a mixture is of interest, 20 °C below its boiling point.
Sample size:	0.1 μl.
Attenuation:	depends on sensitivity of detector and manufacturer's design for amplifier. Somewhere near bottom of range for FID and ECD, and top of the range for TCD.
Flow rate:	40 $cm^3 min^{-1}$.
Injection heater:	about 40 °C above column oven temperature.

I then alter these conditions in the light of the results from the first injection, until a satisfactory chromatogram is obtained. The choice of the stationary phase, the loading and the support were discussed in Section 3.3 of this Unit. What we need to do now is to consider each of the other parameters in turn, to see how it might be altered and what improvement is likely to result.

I have tried to illustrate the effects we shall discuss with chromatograms of a mixture of benzene and cyclohexane, generally on silicone oil columns. On this stationary phase, the separation of the mixture is only partial, so that quite small differences in resolution show up as quite a noticeable difference in the size of the 'valley' between the peaks. The chromatograms were all obtained using a Pye 104 gas chromatograph fitted with a TCD (katharometer).

4.2. LENGTH OF THE COLUMN

Increasing the length of the column increases retention times proportionately, and so increases the difference between them. You can therefore improve resolution just by using a longer column. Unfortunately, the improvement is not proportional to the increase in length. True, if you double the column length, you double the retention times and so double the difference between them. But, because the components take twice as long to pass through the column, they spread more, roughly equal to the square root of the increase in time. Our double length column would therefore lead to peaks 1.4 times as wide, which would detract from the improvement due to the increased difference between retention times. The resolution would improve by a factor of only about 1.4.

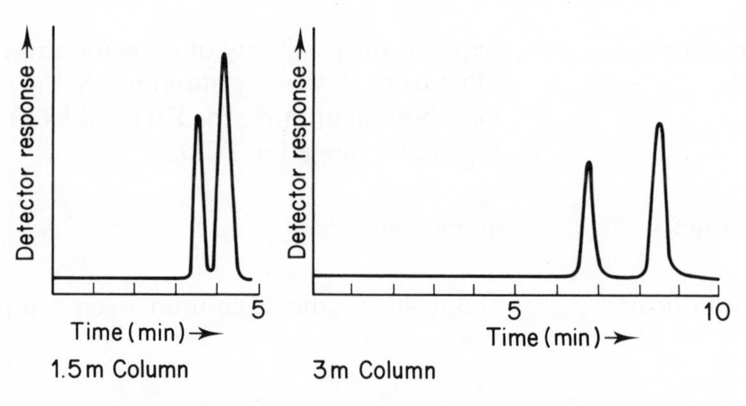

Fig. 4.2. *The separation of benzene and cyclohexane at 60 °C on 10% squalane columns of different lengths*

Still, it can be a significant improvement (see Fig. 4.2) and one which is worth considering if you need the extra resolution for an analysis that is going to be repeated routinely. It would be unrealistic to think in terms of keeping a stock of columns of all lengths in each stationary phase, just in case you might need them for an occasional *one off* analysis, but for a routine analysis, it would be worth packing a special longer column. For a *one off* analysis, it might be better to try something less time consuming, if a little less effective.

Π What is the longest packed column you are likely to be able to use for this purpose?

How much had you remembered of Section 3.2? 10 m was said there to be the maximum length which could be conveniently used. If you had forgotten this perhaps a quick re-read of 3.2 would be a good idea.

Using a shorter column would reduce retention times and lead to a loss of resolution. There will be occasions when you can afford to accept this loss of resolution in order to save analysis time. Once more, this approach is more applicable to setting up a routine analysis than a *one off*. For a *one off*, you could probably achieve the same effect by increasing the column temperature, unless, of course, you were already at the upper temperature limit of the stationary phase, or the sample was thermally labile.

4.3. TEMPERATURE

A component's rate of migration is controlled by its distribution equilibrium between the stationary and mobile phases. In glc, the position of this equilibrium is controlled by both the component's solubility in the stationary phase and its saturated vapour pressure. In very approximate terms, the retention time is inversely proportional to the vapour pressure.

The resolution of two components with no difference in solubilities would therefore depend upon the difference in the *reciprocals* of their vapour pressures. Vapour pressures tend to increase fairly steadily with temperature, so the difference between the vapour pressures of two compounds will not alter enormously as the temperature rises. However, the difference between the *reciprocals* of two large numbers is very much less than the difference between the *reciprocals* of two small numbers with a similar increment between them.

Π If you don't believe me, try it:

$1/1 - 1/2 =$

$1/9 - 1/10 =$

This means that as temperature rises the difference between the reciprocals of the vapour pressures of two components decreases dramatically. You can see this happening in the graphs of the vapour pressures and the reciprocals of the vapour pressures of two hypothetical components shown below. As a result of this, so far as separation is concerned any difference between the vapour pressures of two components will become less significant as the temperature is raised.

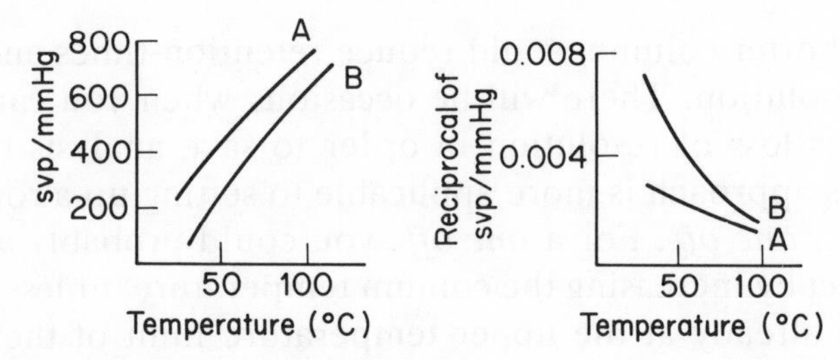

Fig. 4.3a. *Graphs comparing the effect of temperature on differences in vapour pressures and their reciprocals*

The resolution of two components therefore decreases with increasing temperature and increases with decreasing temperature (see Fig. 4.3b). If, therefore, the resolution is not quite good enough on your trial chromatograms you can usually improve it by lowering the temperature. A 10 °C reduction will usually effect a modest improvement, but unless I am working at a low temperature, I tend to favour jumps of 20 °C so that you can see something fairly dramatic occurring reasonably quickly. You can tell whether it is going to work much sooner and so save a lot of time. You will probably have to go back and fine tune it with smaller changes later on, anyway.

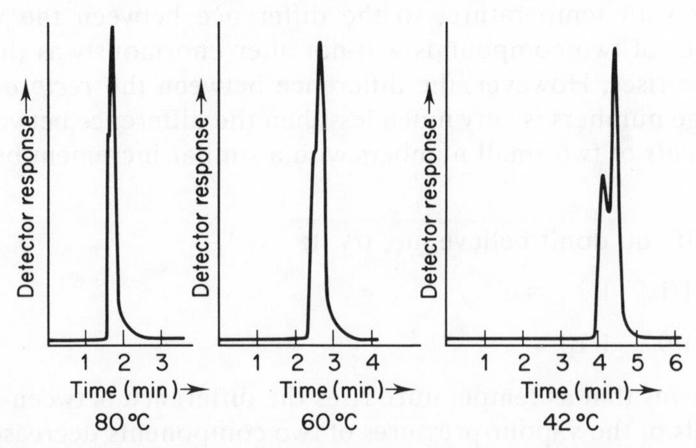

Fig. 4.3b. *The separation of benzene and cyclohexane on a 10% silicone oil column at various temperatures*

Of course, decreasing the temperature increases the retention times and the peak widths, so there is a limit to how far you can go down this road. If the temperature is lowered too much, your peaks become very ill-defined and the analysis time becomes unacceptably long. In practice, I find that it is not often profitable to try to work more than 60 °C below the boiling point of the least volatile component in which I am interested.

Π Study the chromatograms and conditions given in Fig. 4.3b and answer the question 'Would it be reasonable to pursue a policy of lowering the temperature if you wanted to resolve benzene and cyclohexane completely?'

I hope that you agree with me that the answer must be 'NO'. Certainly, lowering the temperature is improving the separation and if the trend continues, lowering it another 40 °C or more should give complete resolution. That would mean operating at 0 °C or less. Gas chromatographs which will do this have been built, but the average laboratory gas chromatograph does not thermostat well below 40 °C, so this is not really a practicable answer.

You can, of course, play the opposite game. If you have plenty of resolution on your trial chromatogram, you can trade some of it off against a saving in analysis time by increasing the temperature. This will reduce both the retention times and the resolution.

4.4. TEMPERATURE PROGRAMMING

Too high a temperature, then, causes poor resolution and carries with it the risks of column bleed and thermal decomposition of the sample. Too low a temperature causes ill-defined peaks and excessive retention times. You would appear to have a problem when you come to analyse a mixture with a very wide range of boiling points. Any temperature which is just right for some components will either be too high or too low for others! The hypothetical gas chromatograms shown in Fig. 4.4a illustrate the sort of thing that you might see.

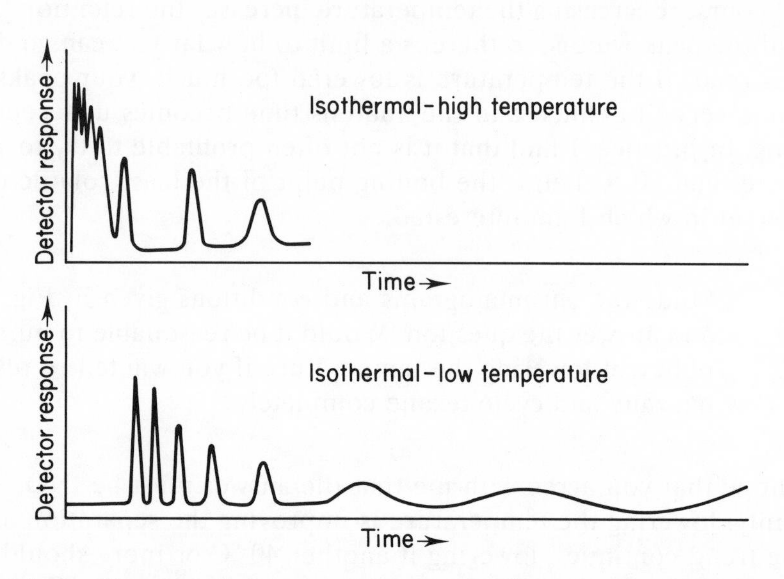

Fig. 4.4a. *Simulated chromatograms to illustrate the resolution of wide boiling range mixtures by isothermal gas chromatography*

The answer to this problem is temperature programming. The analysis is started with a low column oven temperature which is suitable for the more volatile components of the mixture. The temperature is then increased progressively as the analysis proceeds so that it has reached a temperature appropriate to the least volatile components by the end of the analysis. In this way, each component starts to migrate rapidly as the oven temperature reaches a level appropriate for it. Reasonably sharp, well resolved peaks result. (See Fig. 4.4b).

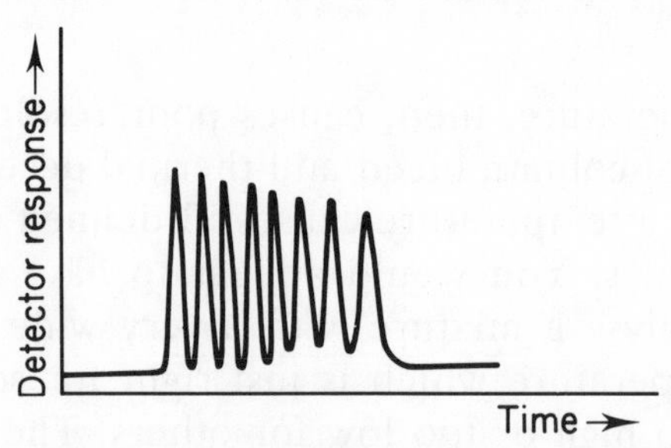

Fig. 4.4b. *Simulated gas chromatogram to illustrate improved separation by temperature programmed gas chromatography*

Π Complete the following statement by selecting appropriate phrases from List A:

'The advantages of temperature programmed gas chromatography may be summarised as follows:

1.

2.

3.

List A

(*a*) Reduced analysis time.

(*b*) Better detection of later peaks.

(*c*) Good potential for analysing mixtures with wide ranges of boiling points.

(*d*) More reproducible retention times.

(*e*) Improved peak shape.

(*f*) Raised upper temperature limit.

I would have chosen (*c*), (*e*) and (*b*). Temperature programming does allow you to analyse 'wide boiling' mixtures, gives overall better shaped peaks and, because later peaks are taller and sharper, gives much better detection of them.

Retention times are likely to be less reliable, because of the difficulty of controlling the temperature whilst it is being increased.

Temperature programming may give you a reduced analysis time, but this is not always the case, and it usually does nothing to raise the temperature limit. (In fact quite the reverse. Since the increase in temperature causes an increase in bleed which can spoil a chromatogram, the demands on the thermal stability of the stationary phase may be greater). These problems are discussed below.

The advantages of this technique may seem obvious. You may safely assume that there will be corresponding disadvantages! Firstly, as the temperature increases during the analysis, so will the stationary phase bleed from the column. This may lead to an exponentially upward drifting baseline and an increase in the noise level if you are working at a low attenuation and approach the upper temperature limit of the stationary phase. Nothing can be done about the noise level, but by fitting a second, identical column and detector into the oven, you can overcome the drift. The two detectors have to be connected so that it is the difference between their signals that is passed to the recorder. In this way the bleed from the second, reference, column approximately cancels out the bleed from the first, analytical, column and a fairly steady baseline results. Computer based data stations or integrators often offer the possibility of storing the baseline from a 'blank run' and then 'subtracting' this from the chromatograms of subsequent actual runs. This does make an assumption about reproducibility between 'runs', but it is another solution to the problem. Secondly, after you have raised the temperature of the column oven during an analysis you will have to wait for it to cool down and restabilise before you can analyse the next sample. This process can take a long time, depending on the temperatures involved. It will certainly add considerably to the analysis time, so I try to avoid this technique unless it is absolutely essential, and stick to isothermal operation.

Most modern gas chromatographs include a facility for temperature programming. It is usually possible to select an initial period at constant temperature followed by a period during which the temperature increases linearly at a preselected rate, and a final period of constant temperature. Such skill as there is lies in selecting the length of each period and the rate of temperature increase. In general, I would caution you against using very high rates of temperature increase. The transfer of heat to the centre of the packed column is not a fast process and the temperature of the stationary phase will lag a long way behind the air temperature of the oven if a high rate is used. I have found it realistic to regard 12 °C min^{-1} as an absolute maximum, though I would prefer not to work at more than 5 °C min^{-1}. This is not, perhaps, quite such a serious problem with capillary columns where the heat does not have to be transferred so far to reach all of the stationary phase.

4.5. SAMPLE SIZE

Ideally, you want the smallest sample size which is easily detectable. Large samples occupy more space as vapour and take marginally longer to vaporise. They start off as longer vapour plugs at the 'top' of the column and give broader peaks than smaller samples. This means slightly poorer resolution, but the effect is not all that significant. Some idea of the scale of the effect can be gained by studying Fig. 4.5a.

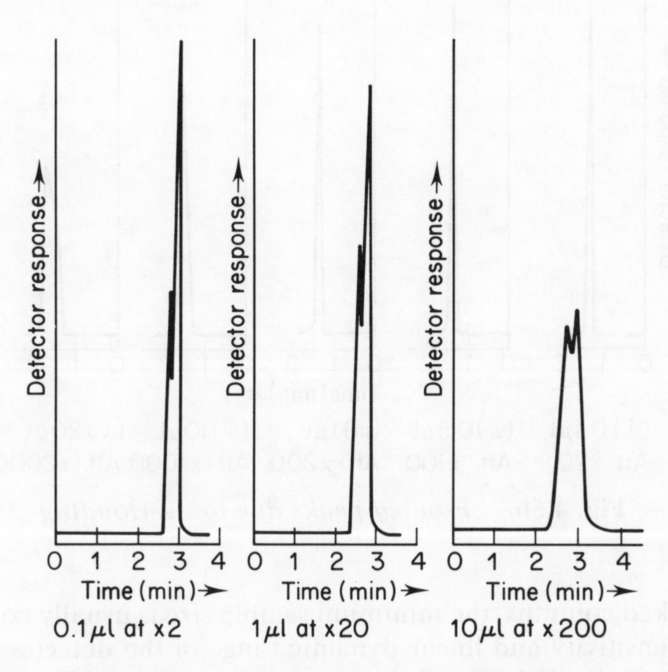

Fig. 4.5a. *Separation of benzene and cyclohexane on a 19% silicone oil column at 60 °C, using various sample sizes and attenuations*

For packed columns this is rarely a limiting consideration below a sample size of 10 μl of liquid. If, however, very large samples are used in glc, there is a noticeable tendency for peaks to 'front'. This is the opposite of 'tailing', and the resulting peaks with sloping front profiles and normal rear profiles are obviously much broader and less well resolved (see Fig. 4.5b).

For capillary columns, however, the problem of overloading is a much more serious one and has lead to the development of stream splitters, as we saw in Section 2.5 of this Unit.

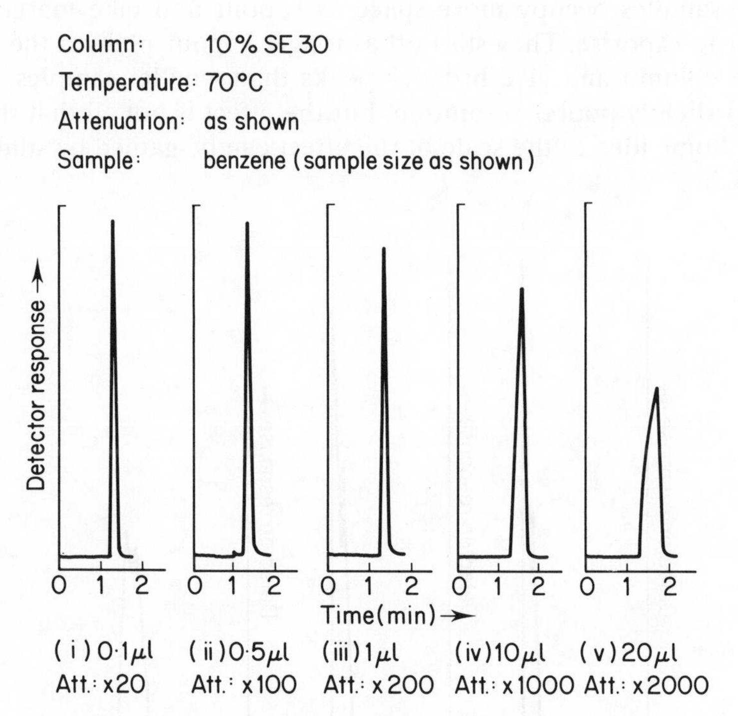

Fig. 4.5b. *Fronted peaks due to overloading*

For packed columns, the minimum sample size is usually controlled by the sensitivity and linear dynamic range of the detector and the accuracy and reliability of the microsyringe being used for injection. This means that for pure liquids and simple liquid mixtures we are talking about 0.1 μl samples for FIDs and ECDs, whilst the TCD (katharometer) is better with larger samples (around 1 μl). Gas samples are proportionately larger (around 1 cm^3). Gas samples of more than 5 cm^3 take a long time to inject and result in very broad peaks, so this probably represents the limit for a packed column.

One result of this upper limit on sample size is that trace analysis is hampered. As you try to determine ever lower concentrations of a trace analyte, the limit of detection of the detector confronts you with a barrier which you can only get round only by using a larger sample, so that there is a larger amount of analyte to detect. However, once your sample size reaches the maximum that your system can handle you will have reached the limit of this approach and with it, the limit of your trace analysis.

There is also another interesting problem if you have badly tailed peaks. Reducing the sample size of very small samples reduces the overall size of the peak without reducing the size of the tail (see Fig. 4.5c).

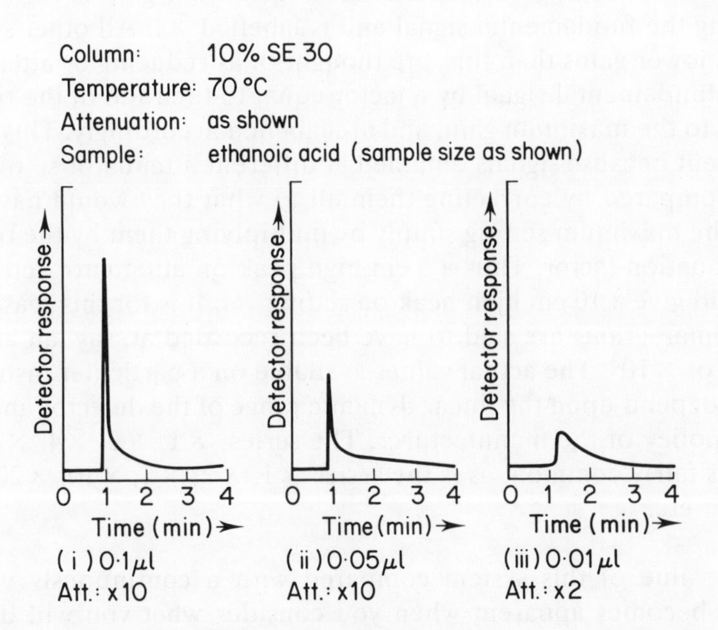

Fig. 4.5c. *The effect of reducing sample size on a badly tailed peak from a very small sample*

This may lead to non-linear calibration graphs. It also means that when the sample gets too small the peak shape and consequent resolution from neighbouring peaks becomes unacceptable. Paradoxically, this is one situation where a larger sample will give better results, since the tail then represents a smaller proportion of the peak. If the attenuation is increased, as it will be to keep the larger peak 'on scale', the tail becomes much less noticeable.

4.6. ATTENUATION

The signal from the detector is always amplified before it is passed on to a chart recorder. In order to ensure that signals from different sized samples are large enough to appear on the chart but not so large as to go 'off scale', the 'gain' of the amplifier must be variable. It is normal for this to be done in a series of discrete steps rather than by a continuously variable control.

The control on the amplifier which performs this is called the *attenuator*. The setting corresponding to maximum gain is regarded as giving the fundamental signal and is labelled $\times 1$. All other settings, with lower gains than this, are thought of as reducing or attenuating this fundamental signal by a factor equal to the ratio of the relevant gain to the maximum gain, and are labelled accordingly. This is convenient because signals obtained at different attenuator settings can be compared by correcting them all to what they would have been on the maximum setting simply by multiplying them by the relevant attenuation factor. Thus a 5 cm high peak on attenuator setting $\times 2$ would give a 10 cm high peak on setting $\times 1$. It is for this reason that chromatograms are said to have been recorded at, say, an attenuation of $\times 10^3$. The actual values available on a particular instrument will depend upon the linear dynamic range of the detector and upon the policy of the manufacturer. The series, $\times 1$, $\times 2$, $\times 4$, $\times 8$, $\times 16$ etc is fairly common, as is the series $\times 1$, $\times 2$, $\times 5$, $\times 10$, $\times 20$, $\times 50$, $\times 10^2$ etc.

The value of this system compared with a continuously variable gain becomes apparent when you consider what you will do after examining your first trial chromatogram. If the peaks are only 1/10th the size you want, you decrease the attenuation by a factor of 10. If they go 'off scale' you increase the attenuation. If you decide to decrease the sample size by a factor of 5 but still want the same sized peaks, you decrease the attenuation by 5.

Π A 0.25 μl sample of benzene, injected into a gas chromatograph at an attenuation of 2×10^5 gave a peak 12.3 cm high.

1. How high a peak would it have given at an attenuation of 5×10^5?

2. How high a peak would it have given at an attenuation of $\times 10^3$? (Assuming that the chart paper was wide enough!).

3. If the chart paper was only 25 cm wide, what size sample would you have to inject at an attenuation of $\times 10^3$ for the peak to stay on scale?

1. 4.92 cm. You could have arrived at this answer by the pedantically correct method:

0.25 μl benzene at $\times 2 \times 10^5$ gave a peak 12.3 cm high

0.25 μl benzene at $\times 1$ would give a peak $12.3 \times 2 \times 10^5$ high

0.25 μl benzene at $\times 5 \times 10^5$ would give a peak $(12.3 \times 2 \times 10^5)/(5 \times 10^5)$, ie 4.92 cm

or, you could have used the common sense short cut:

2×10^5 is 2.5 as sensitive as 5×10^5, so the peak height must be divided by 2.5.

2. 2460 cm. Again, either way would have worked. $\times 10^3$ is 200 times more sensitive than 2×10^5, or divide the calculated peak height at $\times 1$ by 10^3.

3. 0.0025 μl, which is impossible of course. The only way to do it would be to dilute the benzene quantitatively with a suitable solvent and inject a realistic sized sample of the solution, eg a 0.25 μl injection of a 1% solution.

You can arrive at the sample size of 0.0025 μl by looking at the peak height given by 0.25 μl, deciding that it is 100 times too large to fit on the chart paper and dividing the sample size by 100.

However, there will be one problem. If you decrease the attenuation, you will increase the amplification of any baseline noise and drift that is present (see Fig. 4.6, which shows recordings on the baseline obtained at various attenuator settings under otherwise identical conditions. Notice the much greater noise and drift at $\times 1$). There will come a point at which these are no longer acceptable, I would suggest about $\times 10^2$ in Fig. 4.6, and it will be pointless to decrease the attenuation any further. You will then have reached the maximum sensitivity of your system in its current form.

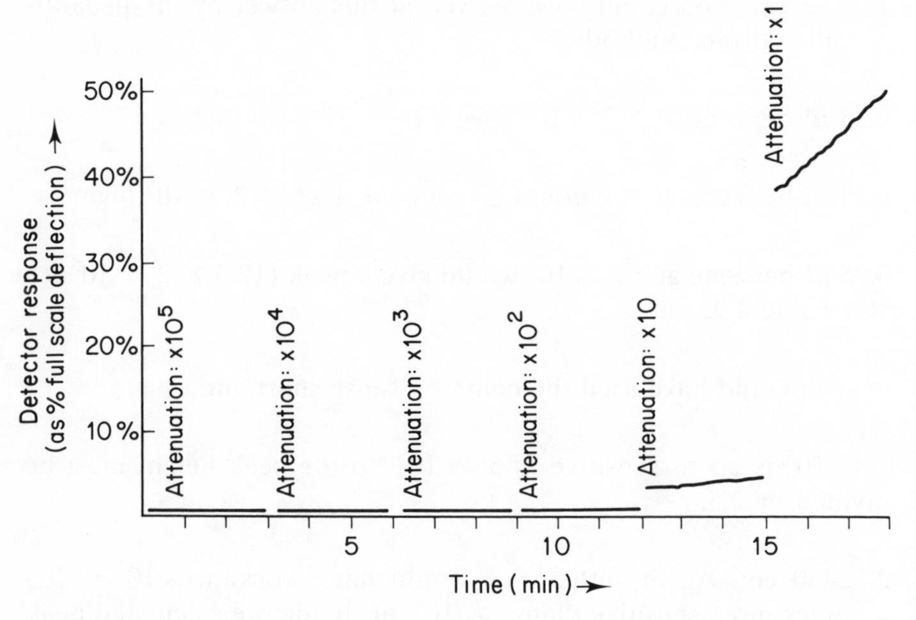

Fig. 4.6. *Baseline noise and drift at various attenuator settings*

There is perhaps one other warning which should be made. Manufacturers use the best available components for building their attenuators, but even so the attenuation factors are seldom more accurate than 2% on a routine instrument. If, therefore, you are likely to compare peaks on different attenuator settings and need a higher precision than 2%, you should calibrate your attenuator. You can do this by connecting an accurate digital voltmeter across the recorder terminals, setting the recorder to 100% full scale deflection, reading the voltmeter, increasing the attenuation and reading it again. The ratio of the voltmeter readings will give the true ratio of two attenuations.

4.7. FLOW RATE

In ACOL: *Chromatographic Separations*, it is explained that both the extent of zone spreading and the column efficiency (the normal measure of which is HETP) vary with flow rate for a number of reasons, but that for any given system there will be a flow rate at which spreading is a minimum and the efficiency a maximum. This is the so called 'optimum flow rate'. For 4 mm id columns the optimum flow rate is around 40 $cm^3\ min^{-1}$ to 50 $cm^3\ min^{-1}$. For narrower or wider bore columns, the optimum flow rate will be smaller or larger, in proportion to their cross sectional areas.

Π What flow rate would you choose to use with a 2 mm id column?

10 to 12 $cm^3\ min^{-1}$. If the flow rate for a 4mm id column is 40 to 50 $cm^3\ min^{-1}$, then that for a 2 mm id column will be less by a factor of $4^2/2^2$, ie by a factor of 4.

If you look back at the conditions for the trial run in Section 4.1, you will see that the flow rate was 40 $cm^3\ min^{-1}$. We are not going to be able to improve the resolution by altering this. What we can do is to save analysis time. If we have an analysis with more than adequate resolution but with very long retention times, we can reduce the retention times by increasing the flow rate and sacrificing a little of our resolution. In fact, the loss of resolution will not be great if we keep the increase in flow rate within reason, since the graph of plate height (as a measure of column efficiency) against flow rate is a fairly shallow curve in the area of its minimum.

If you look at Fig. 4.7 you can see that more than doubling the flow rate to 90 $cm^3\ min^{-1}$ has only caused the previously marginally resolved benzene peak to become a sloping shoulder. Had the two peaks been completely separated you would not have noticed so small a change in resolution. At the same time, we have halved the retention times. There is a similar loss of resolution on slowing down the flow rate to 6 $cm^3\ min^{-1}$, but I cannot think of any reason why you should want to do that except to show a student that van Deemter was right!

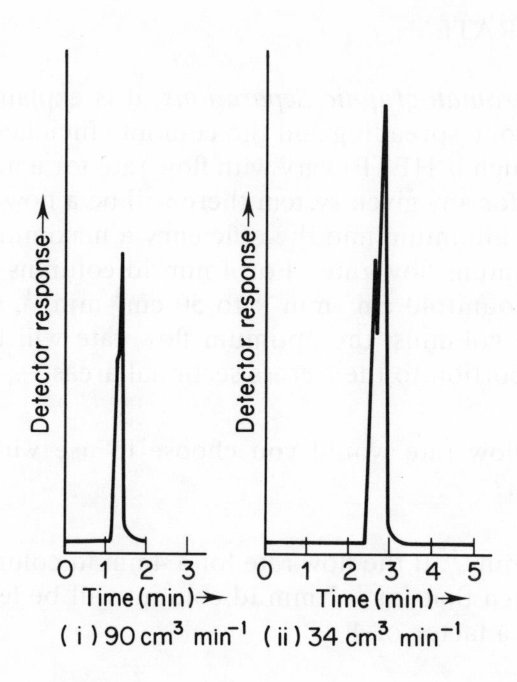

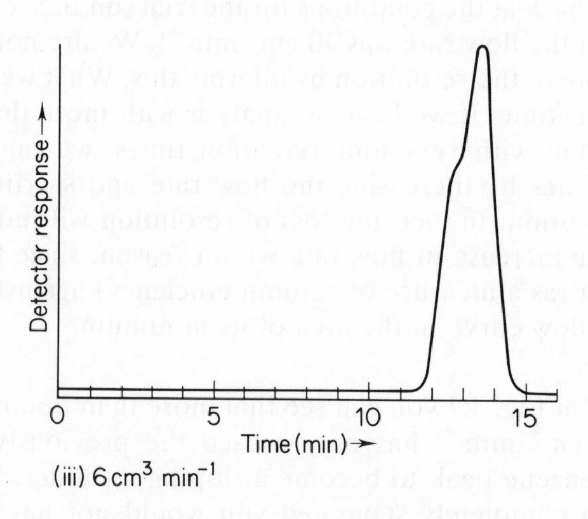

Fig. 4.7. *The separation of benzene and cyclohexane on a 10% silicone oil column at 60 °C showing the effect of flow rate on resolution.*

You are, however, likely to want to increase the flow rate. If you are performing an analysis at or near the upper temperature limit of a stationary phase and have very long retention times, you cannot reduce them by raising the temperature. Increasing the flow rate would be a useful way of reducing the retention times without having to go to the trouble of packing a shorter column.

4.8. INJECTION HEATER TEMPERATURE

The injection heater ensures that the sample is vaporised quickly so that it starts as a narrow plug of vapour. That way, it is more likely to result in narrow peaks and good resolution. You can see the effect of forgetting to turn on the injection heater by looking at Fig. 4.8. The retention times are unchanged but the wider peaks that result are much less well resolved. So, the message is 'Don't forget to switch on the injection heater, and make sure it is hot enough to do its job.'

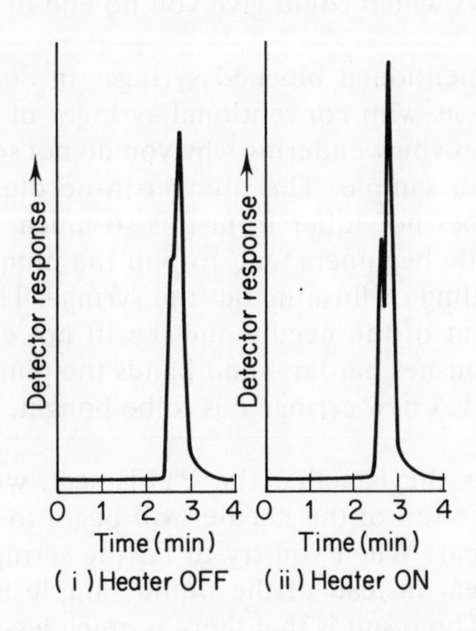

Fig. 4.8. *Separation of benzene and cyclohexane on a 10% silicone oil column at 60 °C, showing the effect of not heating the injection port*

Just as a cautionary tale, though, let me tell you of a student who was doing a project for one of my colleagues. She was trying to determine the ratio of 2-bromo-2-methylpentane and 2-chloro-2-methylpentane in a reaction mixture by glc. She was more than a little perplexed to find more peaks than she expected, even with the pure standards! I solved that one by turning the injection heater DOWN. As the temperature was progressively reduced, the extra peaks disappeared, until by the time the heater was off, we had eliminated the problem. The elevated temperature in the injection zone was causing an elimination reaction, so that injecting an alkyl halide was resulting in two peaks, one for the alkyl halide itself and one for an alkene derived from it. As I have said before, you have to keep your wits about you.

4.9. SO, WHAT ELSE CAN GO WRONG?

We seem to have been talking about an area which borders on trouble shooting, so this is probably a good time to deal with a number of silly little faults which could give you no end of trouble.

I have already mentioned blocked syringes in Section 2.5 of this Unit. This happens with conventional syringes of the type in Fig. 2.5b(*i*), and leaves you wondering why you do not seem to be getting any peaks from a sample. The plunger-in-needle type of syringe (Fig. 2.5b(*ii*)) does not suffer from this so much, but it has other faults. Enthusiastic beginners tend to pull the plunger too far back when they are filling or flushing out the syringe. This pulls the end of the plunger out of the needle and it will not easily go back in again. The user pushes harder – and bends the plunger. Now it will never go back in! A new syringe has to be bought.

Less expensive is the fact that the PTFE seal, which fits around the plunger at the top of the needle, will begin to leak after it has suffered some wear. When you try to fill the syringe, air is sucked down past the seal instead of the liquid sample being sucked up into the needle. The result is that there is much less sample to inject than you anticipate and peaks are very small. Because the leakage is variable, but in general gets worse with continued use, this fault causes successive injections of the same sample to give peaks which are small and variable in size, but tending to get smaller. Once you have recognised it, the answer is to renew the PTFE seal, following the syringe manufacturer's instructions.

Failure to realise that syringes need considerable washing out between samples if 'carryover' is to be avoided is another common error. I usually tell trainees to fill a syringe and then empty it onto a piece of tissue paper at least six times before filling it for injection. If you are not careful about this you get small, extraneous peaks from previous samples turning up in your chromatograph.

You can also get extraneous peaks from a previous sample appearing in your chromatograms if you are impatient. I have seen much time wasted by students injecting a sample, waiting until a few peaks have appeared and the baseline has settled down again and then injecting the next sample. Of course, all the components from the first injection had not come off the column because they had not waited long enough. They turned up in the middle of the peaks from the second injection. Confusion! If you spot it, you have wasted only the amount of time needed to repeat the injections with a greater delay between them. If you do not realise what is happening, you completely misinterpret your chromatogram. The fault ought to be recognisable because the extraneous peaks will be wider than the peaks around them, having been on the column longer. They will also appear to have a variable retention time, since the time between the injection of subsequent samples is likely to vary. It is not always easy to spot it though, and I have been caught out myself. The answer is patience. If in doubt, leave a chromatogram running at least twice as long as you think you ought to!

Another common example of bad technique during injection is to insert the syringe needle into the injection head and then wait for the recorder pen to reach a line on the chart paper before depressing the plunger. The sample vaporises slowly while you are waiting and either broad peaks or double peaks can result. You should depress the plunger immediately the needle has been fully inserted (but see below).

Finally, I have seen trainees either fail to insert the syringe needle fully into the injection port or use a syringe with far too short a needle. In both cases the sample gets injected into the cool region just inside the injection port instead of down in the zone heated by the injection heater. This can result in broad peaks because the sample vaporises slowly, or no peaks at all if it just does not vaporise. Both this fault and the previous one can be avoided if you practice inserting the syringe needle quickly and fully and depressing the plunger promptly. Large samples, if injected quickly, could vaporise more rapidly than they can be carried onto the column. The back pressure might force the vapour plug back through the port, thus contaminating the septum and cooler areas of the carrier gas supply tubing. This is unlikely to be a problem with the small samples usually used. 0.1 μl of benzene will generate about 0.03 cm^3 of vapour, which, at a carrier gas flow rate of 40 cm^3 min^{-1} would take less than one tenth of a second to flush onto the column. However, 10 μl of benzene would be a different matter. It would give 3 cm^3 of vapour which would take 3 or 4 seconds to flush out of the injection port. It might be wiser to depress the syringe plunger at an appropriate rate, so that this vapour is not generated all at once. The problem is likely to be less acute if on-column injection (see Section 2.5 of this Unit) is being used, since much of the sample dissolves in the stationary phase at once, and a smaller volume of vapour is produced.

Most of the other simple faults arise from gas leaks of one sort or another. The continual piercing of the septum by the syringe needle will eventually cause the septum to leak. This in turn leads to a reduced and variable flow rate and correspondingly increased, but variable, retention times. You can usually perform a temporary improvement if you are in a hurry for a result, by tightening down the septum retaining cap. In fact, I am in the habit of wetting my finger and placing it over the hole in the injection head after each injection. I can feel any leaking gas bubbling past my finger and I can then tighten down the retaining cap immediately, before the chromatogram is spoiled. (If you do this, make sure that you don't burn your fingers - the tops of some injection ports get quite hot). The only real answer, though, is to replace the leaky septum. It is really quite a good idea to start each day with a new septum so that this fault is less likely to develop.

The other likely place for a gas leak to develop is between the injection head and the column. With stainless steel columns the seal at this point is usually made by a stainless steel ferrule (olive), and provided you have tightened the fitting adequately, you should have no trouble. With glass columns the seal takes the form of a PTFE ferrule or, in another design, a Viton rubber washer. Both of these can deform or degrade at high temperatures and thereafter cease to do their job. The resulting leak leads to a reduction in flow rate and the usual consequent aberrations. The only answer is to replace the seal.

And lastly, if the flow rate is falling and you cannot find any leaks, it is always worth checking the gas cylinder. It may be nearly empty!

SAQ 4a

A retention time of 2.33 minutes was obtained for ethanol, when it was chromatographed under the following conditions:

Sample size:	0.1 μl
Column:	1.5 m × 4 mm id; 10% PEG 400 on 100–120 mesh diatomaceous earth
Temperature:	70 °C
Flow rate:	40 cm^3 min^{-1}
Injection heater:	90 °C

Select, from the table of retention times (*a*) to (*g*), the value you would expect to get when the above conditions were modified by the changes detailed in (*i*) to (*vi*). In each case all other conditions remain unchanged. ⟶

SAQ 4a (cont.)

(*i*) Flow rate = 55 cm^3 min^{-1}.

(*ii*) Injection heater: off.

(*iii*) Column: 1.5 m × 4 mm id; 5% PEG 400 on 100–120 mesh diatomaceous earth.

(*iv*) Column: 2.7 m × 4 mm id; 10% PEG 400 on 100–120 mesh diatomaceous earth.

(*v*) Column: 1.5 m × 4 mm id; 10% Squalane on 100–120 mesh diatomaceous earth.

(*vi*) Sample size = 1.0 μl.

(*a*) 0.24 min (*b*) 1.28 min (*c*) 1.71 min

(*d*) 2.35 min (*e*) 3.10 min (*f*) 4.23 min

(*g*) 6.05 min

SAQ 4b

The series of chromatograms (*a*) to (*d*) were obtained with successive injections onto a 1.5 m column packed with 10% dinonyl phthalate, at a temperature of 100 °C and a flow rate of 40 cm^3 min^{-1}. The samples were taken from the atmosphere of a badly ventilated paint spraying booth, about which complaints had been made. The analyst's technique is faulty. Can you spot his mistake.

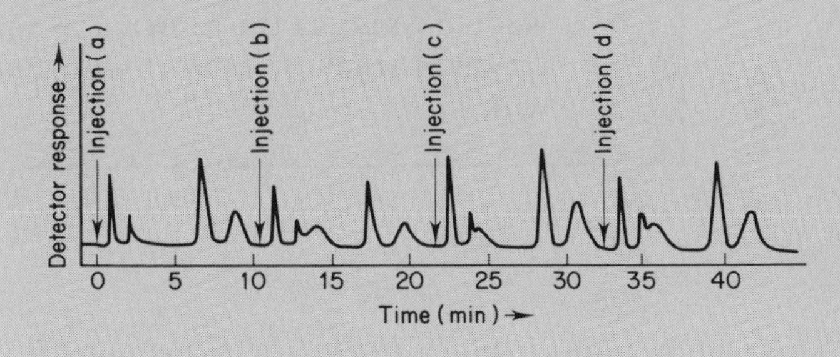

SAQ 4c

A 0.1 μl sample of ethanol injected into a gas chromatograph fitted with a flame ionisation detector, gave a peak 9 mm high at an attenuation of $\times 10^5$

(*i*) How high a peak would be given by 1 μl of an aqueous solution containing 100 mg dm^{-3} at an attenuation of $\times 10^2$?

(*ii*) What attenuation would you use if you wanted to inject 0.1 μl of a 1% aqueous solution of ethanol, if the chart paper is 25 cm wide?

SAQ 4d

The three chromatograms below are the result of three successive, identical injections of the same sample. What has gone wrong, and what should be done to correct it?

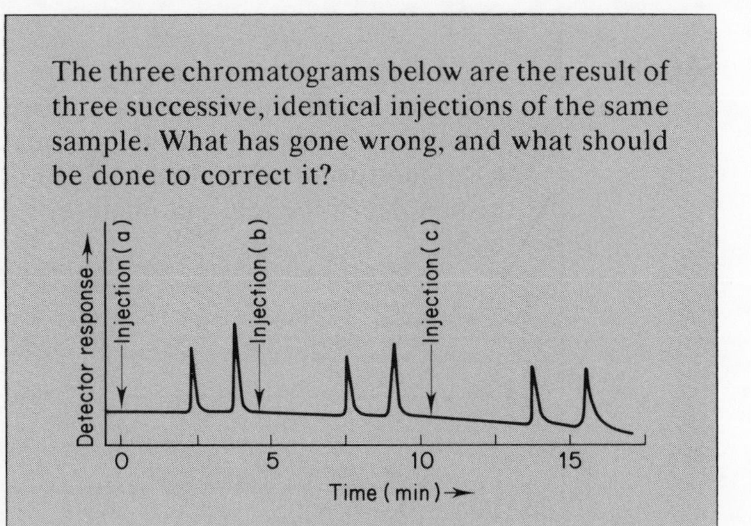

SAQ 4e

The following chromatograms were obtained under the conditions appended to each one. In each case, what changes would you make in order to improve them and make them usable for the analysis of the relevant mixture?

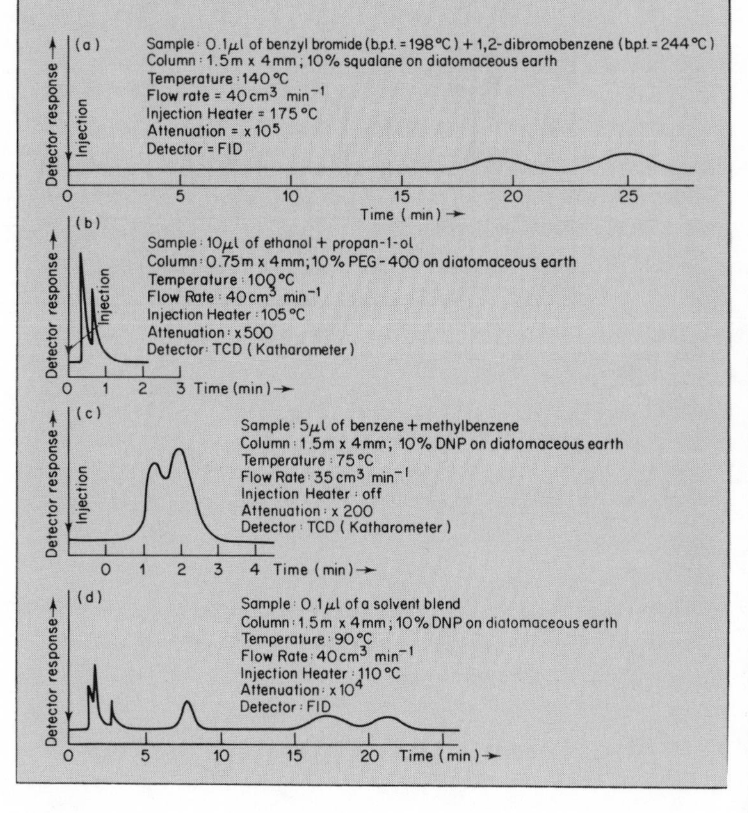

SAQ 4e

Learning Objectives

After studying the material in Part 4, you should now be able to:

- choose the column dimensions, temperature, flow rate, sample size and attenuation appropriate for a given analysis;
- decide when to use temperature programming;
- recognise faults commonly encountered in gas chromatography.

5. The Gas Chromatography of Less Volatile Samples

5.1. INTRODUCTION

At the beginning of Section 3.3 we discussed the 'upper temperature limit' of liquid stationary phases. This is the temperature above which they are unusable because they either vaporise or decompose and so allow too much material to pass into the detector. Then in Section 4.3 we found that a glc column needs to be used at not more than 60 °C below the boiling point of the least volatile component of the sample being analysed. Taken together, these limitations mean that it will be impossible to analyse some not-very-volatile mixtures by glc. There will be no conventional liquid stationary phase capable of being used at a high enough temperature for the mixture's migration to be fast enough to give usably short retention times.

No-one could leave it there. Glc has proved far too useful for us not to try to extend its use as far as possible and much ingenuity has gone into solving the problem posed above. There have been two lines of attack – to extend the temperature range of stationary phases to higher 'upper temperature limits' and to increase the volatility of the mixture being analysed by subjecting it to an appropriate chemical reaction before analysis.

To complete the picture, perhaps we should record that there was, at the same time, a concerted effort to develop liquid chromatography (lc) into an analytical tool similar to gas chromatography. After all, if you have a non-volatile sample to analyse it does seem silly to employ a technique which relies upon volatilisation! The development of lc suffered many delays and disappointments and for a long time the extension of gas chromatography to less volatile samples continued to receive much attention. In the short term, it was the only technique that was promising reasonable results. However, now that lc has reached a much more advanced state, it may be that it has taken some of the thrust out of the development of this area of gas chromatography. Even so, it is still an area which can be profitably explored by the analyst.

5.2. HIGH TEMPERATURE STATIONARY PHASES

The upper temperature limit of a stationary phase is determined by the extent to which it releases volatile material into the gas stream. This material may consist of vaporised molecules of the stationary phase, degradation products of the stationary phase, or impurities which dissolved it. Any attempt to raise the upper temperature limit must seek to reduce such emissions.

Π Suggest one step which you might take to ensure a minimal emission of vaporised stationary phase molecules into the gas stream.

I expect that you suggested using a stationary phase with a lower saturated vapour pressure. You may have taken it further and suggested that a polymer should be used since it will have a very low vapour pressure because of its high relative molecular mass. You may even have taken it further still, and suggested using a specially prepared sample of the polymer with an exceptionally high relative molecular mass. This is indeed one approach, and most of the best conventional high temperature stationary phases are polymers, as you can see from the following list.

Stationary phase type	Example	Temperature limit (°C)
Polyester	PEG-A	190
Polyamide	Versamid 900	350
Polyether	Carbowax 20M	200
Polysiloxane	SE 30	300
	OV 17	350
Hydrocarbon	Apiezon L	300

The only exception is Apiezon L, a high relative molecular mass hydrocarbon fraction isolated from petroleum. It was originally developed as a high vacuum stopcock grease, but has long been used as a stationary phase for glc.

If the problem is the emission of volatile impurities, this could be overcome by their removal. Traditionally this is done by the careful 'conditioning' of columns before use and this involves passing the carrier gas through the column for many hours at a temperature some 25 °C above the upper temperature limit until most of the impurities have been eluted. The column can then be used at temperatures up to the temperature limit with very little increase in baseline level from the detector. Of course, the process of conditioning could be made easier by careful preparation of the stationary phase in the first place, so that there are fewer impurities to remove.

If the problem is caused by the decomposition of the stationary phase, it has to be born in mind that this can be catalysed by small quantities of acids, bases or other reagents present in it. Careful exclusion or removal of such catalysts during manufacture could mean that the onset of decomposition will be delayed until higher temperatures are reached. You could, in this way, raise the upper temperature limit by at least a few degrees.

By paying attention to these points (higher relative molecular mass polymers, use of higher purity materials during manufacture and exclusion or removal of catalysts) at least one chromatography supply house has been able to offer a range of polyester stationary phases with temperature ranges extended by about 20 °C, compared with their commercial rivals.

Π There is one source of catalysis which we have not discussed and which ought to be eliminated if you want to extend the temperature limits of a stationary phase as far as possible. Can you identify it?

If you suggested the support, you have chosen the most obvious answer. Since the only materials that are present, other than the stationary phase, are the sample, the mobile phase, the support, and the column, it has to be one or more of these. You cannot do much about the sample, as a rule. You are generally expected to analyse it 'as received', so if there are unwelcome impurities present you have to put up with them. However, by using a preliminary clean-up procedure, you may be able to ensure that the most obvious potential catalysts are removed. Most carrier gases are pretty pure, especially if there is a molecular sieve trap in the gas line. (Of course, it would be sensible to make sure that there is no oxygen present in the carrier gas, but strictly speaking, oxygen isn't a catalyst, is it?) Any residual acid groups on the support, or indeed any polar groups, could act as catalysts and promote the decomposition of the stationary phase, so you need to remove or block them very carefully. You therefore have to pay particular attention to choosing an inert support and to washing it and silanizing it (see Section 3.4 of this Unit), if you intend to work at very high temperatures. The inside surface of the column is less likely to be a problem (there is less surface area for a start), but it would obviously be a good idea to wash it and silanize it (see Section 3.4 of this Unit) just in case.

As long ago as 1969, P S Wood, by paying attention to all the points so far mentioned, was operating an Apiezon L column up to a temperature of 375 °C, at least for short periods of time. This is some 75 °C above what is normally considered to be the upper temperature limit for this phase.

An alternative approach to solving this problem would be to look for new, more thermally stable, classes of polymer. This is more easily said than done, but some successes have been claimed. One of these was Dexsil, Fig. 5.2, a carborane modified polydimethylsiloxane, claimed to be stable up to 500 °C.

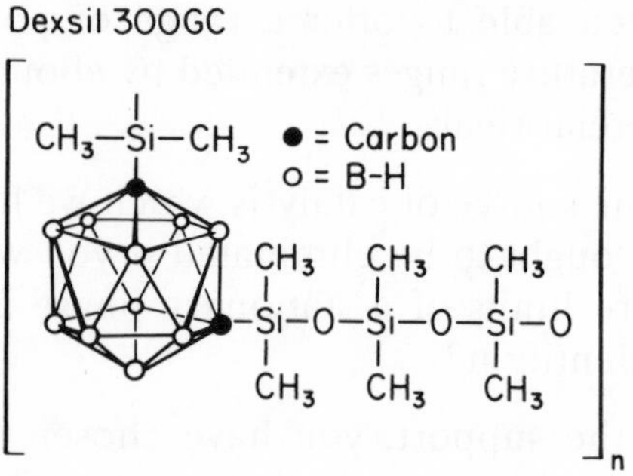

Fig. 5.2. *The structure of Dexsil 300GC*

Another was the development of 'bonded phases'. In these, the liquid stationary phase is chemically bonded onto a silica support, usually by a silyl ether linkage:

$$-\overset{|}{\underset{|}{\mathrm{Si}}}-\mathrm{O}-\mathrm{Si(Me)_2}-\mathrm{R}$$

Such approaches partially overcome the problem of stationary phase volatility and, in theory, the upper temperature limit is set only by the thermal decomposition of the samples involved. These column packings are, however, expensive and the range available is limited. Those available include octadecyl groups bonded to silica (non-polar), polydimethylsiloxane chains bonded to silica (intermediate) and polyesters bonded to silica (very polar). Stationary phase bleed is certainly reduced, but there are problems with capacity. The layer of stationary phase will be one molecule thick, a lot less than a conventional packing. It will be capable of dissolving much less sample. Indeed, there is debate over whether a monomolecular layer of stationary phase can be considered to dissolve anything, or whether components are retained by something more akin to adsorption. After an initial flurry of interest, they have settled down to a modest, rather than extensive, use in glc. They have, however, found much more application in lc. In liquid phase partition chromatography, there is a tendency for the stationary phase to dissolve in the mobile phase and to be stripped from the column. As the detectors used for lc became more sensitive they revealed that such stationary phase bleed was a serious problem. It interfered with detection and shortened column life. Bonded phases overcame this problem, and it is probably here that they have their greatest future.

5.3. DERIVATISATION

The second approach which we proposed for the analysis of relatively involatile mixtures was to increase their volatility.

∏ Study List A and then select from List B the factor which has the most striking effect on boiling point.

List A

Compound	bp (°C)
H_2O	100
CH_3OH	68
$CH_3CO_2CH_3$	57
CH_3OCH_3	−24

List B

1. The number of atoms in the molecules

2. The number of hydrogen atoms attached to an electronegative atom

3. The relative molecular mass of the molecules.

If you thought the answer was:

1. the number of atoms in the molecule, how did you account for dimethyl ether having the lowest boiling point whilst having an intermediate number of atoms?

2. the number of hydrogen atoms attached to an electronegative atom – well done! I would agree with you, for these compounds. This is because such hydrogen atoms can form intermolecular hydrogen bonds, so that more energy is needed to separate the molecules in changing the substance from a liquid into a vapour.

3. the relative molecular mass, you have noticed a minor effect, although it is heavily masked in these examples by the hydrogen bonding described above. You would expect molecules of higher relative molecular mass to require a higher temperature to give them enough kinetic energy to escape from the liquid as vapour (to boil). The series water–methanol–dimethyl ether goes in the reverse order because of the decreasing hydrogen bonding, which outweighs the increasing relative molecular mass. The effect of relative molecular mass appears to be noticeable when you compare dimethyl ether and methyl methanoate, neither of which can hydrogen bond. Even here it is complicated by the fact that methyl methanoate is more polar because of the carbonyl group. The larger dipole–dipole attraction within the liquid will lead to more energy being needed for vaporisation, and consequently a higher boiling point.

In many classes of compound, then, intermolecular hydrogen bonding is a major cause of high boiling points. Its prevention would eliminate a significant intermolecular attraction and consequently increase volatility. Such an approach might well be useful in analysing high boiling alcohols, phenols, carboxylic acids and amines. There is also likely to be an added bonus. Hydrogen bonding between the analyte molecules and polar groups on the support causes serious tailing in glc. If we prevent the analyte molecules from hydrogen bonding, we shall reduce the tailing which is such a feature of the gas chromatograms of alcohols, phenols, carboxylic acids and amines.

Prevention would seem to be a matter of replacing the offending hydrogen atoms by forming a suitable derivative, which would then have a lower boiling point and might be more amenable to glc. The reaction in which the derivative is formed would have to be convenient and fairly easily carried out. It would need to be reliable, cause no side reactions which might give rise to spurious peaks, and, if quantitative analyses were to be attempted, it should take place in as near to 100% yield as possible.

Whilst you might quite correctly conclude from the earlier question that esters are less effective than ethers at lowering boiling points, their preparation is much easier and considerably more reliable. They are therefore more likely to be satisfactory as derivatives for glc. Many alcohols, phenols and amines have been successfully gas chromatographed as their acetyl derivatives. They are usually formed by reacting a small amount of the mixture to be analysed with ethanoic anhydride in the presence of a catalyst (often hydrochloric acid) for about an hour at 100 °C.

$$ArOH + (CH_3CO)_2O \rightarrow CH_3CO_2Ar + CH_3COOH$$

This is conveniently performed by enclosing the reactants in a small sealed vessel placed in an oven or water bath. A range of screw capped vials has been produced for this purpose. The effectiveness of this approach may be judged by examining the chromatograms of phenol and its acetyl derivative in Fig. 5.3a (notice the improvement in peak shape, but lack of reduction in retention time in this example).

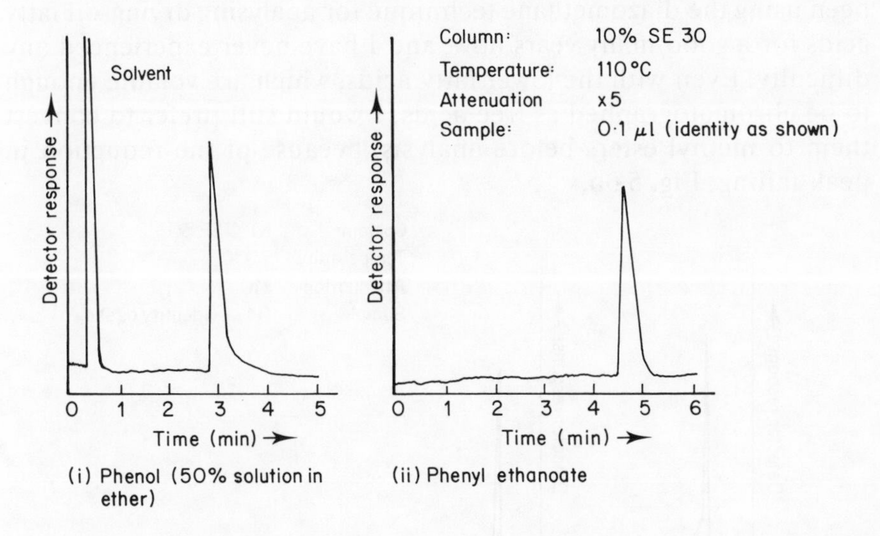

Fig. 5.3a. *Chromatograms to illustrate the improvement in peak shape on conversion of a phenol into an ester*

A further development of this approach has been the use of trifluoroacetyl derivatives. These are somewhat more volatile than the acetyl derivatives and offer greater sensitivity if an electron capture detector is being used (see Subsection 2.6.3 of this Unit). They do, however, require more expensive reagents for their preparation. They can be prepared by heating the sample to be analysed with trifluoroethanoic anhydride or N-trifluoroacetylimidazole in a screw capped vial at 60 °C, for about an hour.

$$RNH_2 + (CF_3CO)_2O \rightarrow RNHCOCF_3 + CF_3COOH$$

$$RNH_2 + CF_3CON \quad N \rightarrow RNHCOCF_3 + NH \quad N$$

Carboxylic acids are invariably chromatographed as their methyl esters. Various ways are available for their preparation, but treatment at room temperature with an ethereal solution of diazomethane or with a methanolic solution of boron trifluoride are probably the most convenient. Even here, where very mild conditions are used, there has been some debate about the possibility of side reactions and the extent to which 100% conversion can be relied upon. I have been using the diazomethane technique for analysing drying-oil fatty acids for a good many years now, and I have never experienced any difficulty. Even with the lower fatty acids, which are volatile enough to be chromatographed as free acids, I would still prefer to convert them to methyl esters before analysis, because of the reduction in peak tailing, Fig. 5.3b.

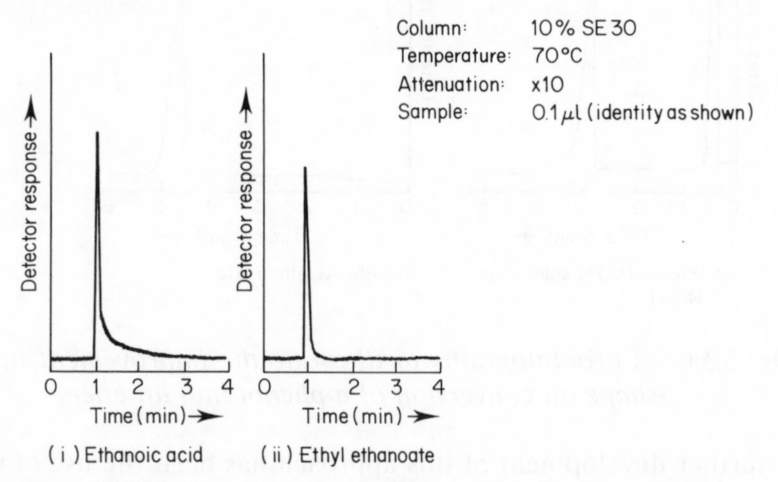

Fig. 5.3b. *Chromatograms to illustrate the improvements in volatility and peak shape on conversion of an alkanoic acid into an ester*

A later development was the use of trimethylsilyl derivatives. These can be prepared from alcohols, phenols, amines or acids quite quickly, often at moderate temperatures, and they are usually more volatile than simple esters or amides. It is even possible to gas chromatograph such unlikely materials as carbohydrates and amino-acids by means of them. The earliest method of preparing trimethylsilyl derivatives was by reaction with trimethylchlorosilane (TMS – Me_3SiCl), often in pyridine solution:

$$ROH + Me_3SiCl \rightarrow ROSiMe_3 + HCl$$

Hexamethyldisilazane (HMDS – $Me_3Si{-}N{=}N{-}SiMe_3$) was later introduced as an improved reagent, but it is often used as a mixture with TMS. This is a particularly powerful reagent. Neither reagent is easy to store and for this reason a later introduction, N,O-bistrimethylsilylacetamide (BSA – $CH_3C(OSiMe_3){=}NSiMe_3$), which is both easier to store and use, is now probably the reagent of choice.

$$2\,ROH + CH_3\overset{\overset{\displaystyle OSiMe_3}{|}}{C}{=}N{-}SiMe_3 \rightarrow 2\,ROSiMe_3 + CH_3CONH_2$$

Silyl derivatives are quite easily made by reacting the analyte with one of the silylating agents in a screw capped vial, either at room temperature or at about 60 °C. They offer a very convenient way of gas chromatographing compounds which would otherwise be too involatile due to hydrogen bonding. Their main disadvantage is the possibility of forming volatile reaction products, such as hexamethyldisiloxane and N-trimethylsilylacetamide, which can result in spurious peaks in your chromatograms.

5.4. PYROLYSIS GAS CHROMATOGRAPHY

There are some compounds that are involatile, not because of hydrogen bonding, but because their relative molecular masses are so great that dispersion forces between the molecules in the liquid phase are very large. The molecules cannot then acquire the necessary energy to overcome these forces at any realistic temperature.

The formation of derivatives cannot help to overcome this problem and make the compounds amenable to glc. The only solution would seem to be to break the large molecules up into smaller fragments which would be more volatile and so amenable to gas chromatography.

This approach might be valuable for identifying and quantifying non-volatile compounds by glc if:

1. The fragments produced are characteristic of the compound.

2. The reaction in which they are formed is reproducible.

3. The whole process is reasonably quick and simple to perform.

There are many reactions that can be used to fragment large molecules, such as hydrolysis, ammonolysis, oxidation and pyrolysis. Pyrolysis is the most useful of them. Hydrolysis is often slow, the initial reaction mixture has to be purified and dried before chromatography, and the products will usually be prone to hydrogen bonding. Ammonolysis is not generally satisfactory for similar reasons. Oxidation, if carried out with a chemical oxidising agent, will also give a wet, contaminated product, and whilst catalytic oxidation with molecular oxygen might be more convenient, some oxygen would inevitably reach the column when the volatile oxidation products were transferred straight onto it. We have seen already that this is undesirable.

Pyrolysis avoids most of these difficulties. Reagents, which might cause difficulties if they are not removed, are not added. There is no need for lengthy and involved clean up procedures, and it will be easy to transfer the volatile pyrolysis products direct to the column. Photolysis might enjoy similar advantages.

The technique has become known as *pyrolysis gas chromatography*, or pgc for short. Its history has been one of steady improvements in satisfying the three conditions – especially the second (reproducibility).

This is not to say that the other reactions cannot be used. Indeed, a preliminary hydrolysis, followed by a clean up, is a standard way of preparing glycerides for gas chromatography. It is just that pyrolysis will be much simpler and quicker if it is appropriate for the materials of interest.

Pyrolysis gas chromatography (pgc) will not, of course, be equally applicable to all non-volatile compounds. The chromatograms will be difficult to interpret if they consist of a very large number of different sized peaks. It follows that we shall find pgc of most value for analysing compounds which, on pyrolysis, produce as few fragments as possible – preferably only one! In practice, it turns out that the most useful work has been done in the field of polymer analysis – especially the analysis of addition polymers. Many of these depolymerise quite easily on modest heating. There is limited side reaction and it is not unusual for a pure polymer to give a chromatogram consisting of a single large peak due to the monomer, with just a few, much smaller, peaks due to dimers and degradation or rearrangement products. You can see this pattern in the chromatograms given by polymethyl methacrylate, polystyrene and polyethylene shown in Fig. 5.4a.

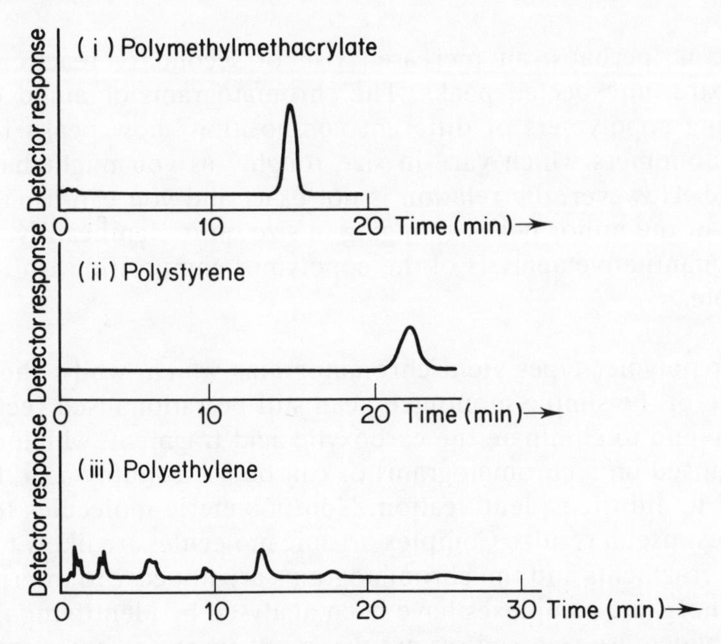

Fig. 5.4a. *Typical chromatograms obtained by PGC, using a Porapack Q column, temperature programmed 100 °C to 200 °C at 8 °C min^{-1}*

Copolymers, not surprisingly, give peaks for both monomers, as you can see from Fig. 5.4b, obtained by pyrolysing a copolymer of styrene and methyl methacrylate.

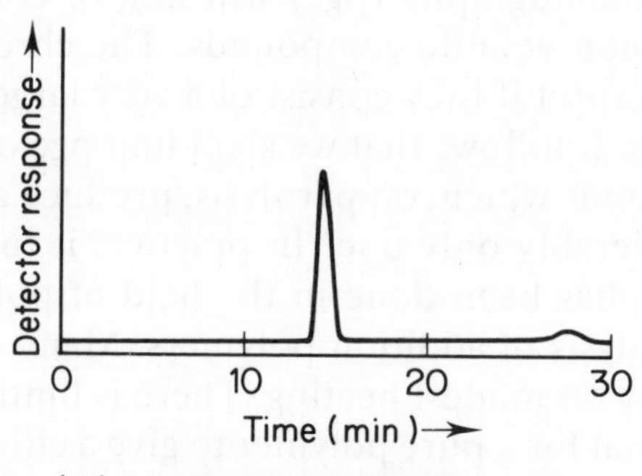

Fig. 5.4b. *Chromatogram obtained by the PGC of a copolymer using a Porapack Q column, temperature programmed 100 °C to 200 °C at 8 °C min^{-1}.*

There is, perhaps, an increased risk of secondary reactions giving extra, unexpected peaks. The chromatograms obtained by pyrolysing copolymers of different composition show peaks for the two monomers which vary in size roughly as you might have expected. However, the relation is not exact and you can usually see some of the minor peaks changing in size quite significantly. Accurate quantitative analysis of the copolymer composition isn't really feasible.

Other polymer types yield chromatograms which, whilst they may not be of the simple monomer , can still be rationalised (cellulose esters tend to eliminate the carboxylic acid fragment, which can be recognised on a chromatogram) or can be used simply as a 'fingerprint' to aid their identification. Non-polymeric molecules tend to give less useful results. Complex organic molecules are likely to yield many fragments and the chromatograms are not easy to interpret or use. Inorganic complexes have been analysed by identifying the ligands which are released on pyrolysis, but there are other, perhaps better, ways of analysing such compounds. Perhaps the most bizarre application is the 'typing' of bacterial cultures by pgc. On pyrolysis, different bacteria produce distinct and recognisable chromatograms which may be used to identify them quite reliably.

The equipment used for this technique is usually referred to as 'a pyrolyser'. It is fitted directly onto the top of the column in place of the injection head, so that pyrolysis can be carried out and the products of reaction swept by the carrier gas straight onto the column for analysis. In this way the whole operation is simple to conduct, so satisfying our third requirement.

To be satisfactory, a pyrolyser must be able to:

1. Control accurately the temperature at which pyrolysis takes place.

2. Raise the temperature of the sample to the pyrolysis temperature quickly, and always at the same rate.

3. Sweep the volatile pyrolysis products onto the column before secondary reactions can occur.

Broadly speaking, pyrolysers fall into three groups: furnaces, flash heaters and Curie-point pyrolysers, none of which satisfies all of the above conditions.

Furnaces consist of a small, heated enclosure through which the carrier gas is made to flow before passing into the column. They include in their design an arrangement for inserting the sample that is to be pyrolysed into the heated enclosure, without interrupting the flow of the carrier gas. In use, the furnace is heated to a pre-selected temperature which is maintained for the duration of the experimental work, the sample is transferred to the heated zone and the products of pyrolysis are swept onto the column for analysis. This type of pyrolyser allows a very accurate control and measurement of pyrolysis temperature, but it does not ensure a rapid and reproducible heating of the sample, and pyrolysis continues indefinitely until there is no sample left. Secondary reactions leading to confusing chromatograms could well be a serious problem under these circumstances.

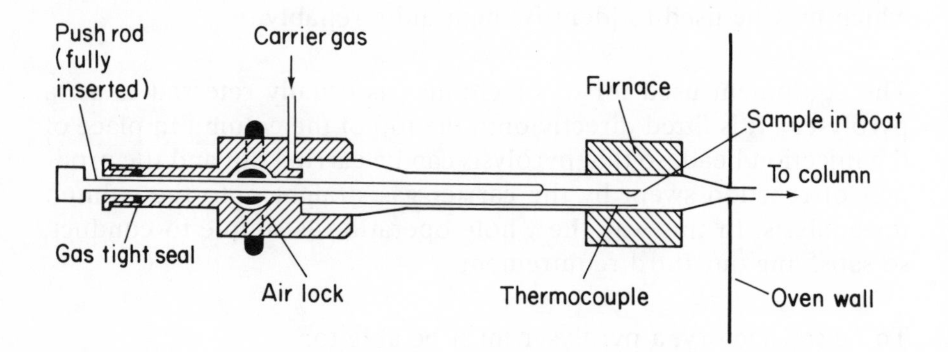

Fig. 5.4c. *A furnace pyrolyser*

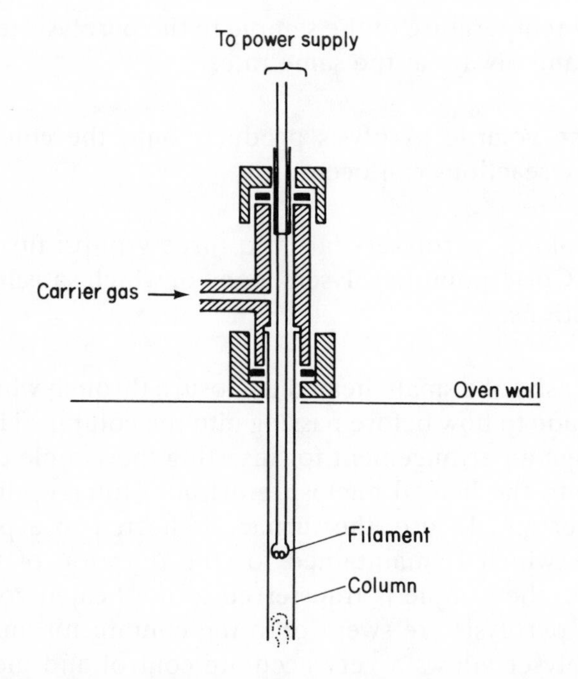

Fig. 5.4d. *A flash heater or filament pyrolyser*

A flash heater is constructed so that the sample can be deposited on a resistance heater in the form of a wire, ribbon of metal or metal boat. A current is passed through the heater, raising it quickly to a temperature that is determined by the current and by heat losses to the carrier gas. The current is usually supplied by the discharge of a capacitor, so there is some control over both the temperature and the duration of pyrolysis. The control is by no means precise and is seriously affected by changes in the flow rate of the carrier gas.

Flash heaters, then, offer rapid heating, short pyrolysis times and rapid cooling, but all at the expense of accurate control over the pyrolysis temperature and accurate knowledge of what is is.

Curie-point pyrolysers, when they were introduced, were going to change all this. To understand how a Curie-point pyrolyser works you need first to know that the paramagnetism possessed by certain pure metals and alloys disappears suddenly if their temperature is raised above a certain limiting value (the Curie-point for that metal or alloy) and reappears again as soon as the temperature falls below that value.

When such paramagnetic metals are placed in a fluctuating electromagnetic field, eddy currents are induced in them. These currents will heat up the metal, and if the field is strong enough, the temperature of the metal will rise until it reaches the Curie-point. The disappearance of the metal's paramagnetism at this temperature means that eddy currents can no longer be induced in it, and the metal ceases to heat up. Any tendency to cool down results in the eddy currents beginning to flow again as the metal's paramagnetism returns, and the temperature is maintained at the Curie-point. A pyrolyser based upon this principle can be fairly simple. A short, quartz tube, through which the carrier gas flows, is attached to the top of the column. An induction coil is placed around this tube and a wire made from a metal with a suitable Curie-point is fitted into the tube. A power supply capable of generating a powerful radio frequency signal is attached to the coil.

Activating the power supply for a pre-selected period of time rapidly raises the temperature of the wire to its Curie-point, maintains it there for the duration of the time period and then allows it to cool down rapidly as the power supply is turned off. If a sample has been

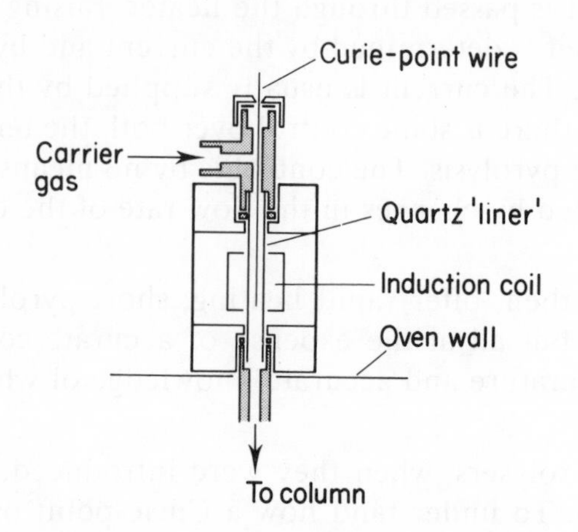

Fig. 5.4e. *A Curie-point pyrolyser*

deposited on the surface of the wire, it should follow the same temperature cycle, and enjoy ideal conditions for pyrolysis. Different pyrolysis temperatures can be obtained by using wires of different alloys with known Curie-points.

Of course, it does not work quite that well! There is some evidence to show that cooling by the carrier gas prevents the surface of the wire reaching its Curie-point, and there are difficulties with higher boiling pyrolysis products condensing on the cool quartz tube before they get to the column. Nonetheless, Curie-point pyrolysers work quite well and are convenient to use. I have found one to be very useful for identifying the polymers used in the surface coatings and plastics industries.

Summary

The value of gas chromatography is such that there will always be pressure to extend its use to ever less volatile compounds. This has been approached in three ways:

1. The use of special stationary phases which have higher temperature limits. This approach has had some success, but there are obviously limits to its development.

2. The formation of more volatile derivatives of the components of mixtures (esters or silyl derivatives). This has been very successful with alcohols, phenols, acids and amines, and with related compounds such as amino-acids and carbohydrates. It has the added advantage of reducing tailing, to which all of these compounds are susceptible.

3. Pyrolysis followed by analysis of the volatile reaction products. This has been most successfully applied to the analysis of polymers and paints, where it has proved a valuable tool.

If you wish to read more about pyrolysis gas chromatography, I would recommend the reviews published by R L Levy and R W May.

SAQ 5a

Would you expect the following analyses to require special treatment because of the low volatility of the sample?

(*i*) The fatty acids in a sample of soap.

Y / N

(*ii*) A sample of diesel oil.

Y / N

(*iii*) The phenols used as raw materials for preparing phenolic resins

Y / N

(*iv*) A phenolic resin.

Y / N

(*v*) A light machine oil.

Y / N

SAQ 5b

Indicate, by circling T for True and F for False, whether you agree with the suggestion that the temperature limit of a column packed with 10% polyethyleneglycol adipate (PEG-A) on diatomaceous earth could be raised by:

1. Adding 0.5% phosphoric acid to the stationary phase.

   T / F

2. Coating the stationary phase onto PTFE beads instead of diatomaceous earth.

   T / F

3. Using carefully washed and silanized diatomaceous earth.

   T / F

4. Passing carrier gas through the column overnight at the temperature limit.

   T / F

SAQ 5c

Complete the following equation:

$$C_6H_5OH + CH_3{-}\underset{}{\overset{\displaystyle OSiMe_3}{\overset{|}{C}}}{=}N{-}SiMe_3 \rightarrow$$

How would you carry out this reaction?

SAQ 5d

Complete the following paragraph by inserting the most appropriate word or phrase, chosen from the list given below, into the blank spaces.

The .... pyrolyser offers the most accurate control of the temperature of pyrolysis, but is probably more prone to .... than most, since it does not control .... On the other hand, the .... pyrolyser offers quite good control of the temperature of pyrolysis, but since it is also possible to achieve .... heating and cooling and .... timing of the duration of pyrolysis, it is probably less prone to secondary reactions.

Cooling	modest
Curie-point	overheating
duration of pyrolysis	secondary reactions
flash heater	slow
furnace	temperature.

SAQ 5e

Select the technique from (*i*) to (*iii*) which you think would be most appropriate for the gas chromatography of the following compounds:

1. D-Mannitol (a polyol found in mushrooms which acts as an antifreeze and prevents frost damage).

2. Perspex.

3. Dinonyl phthalate plasticiser.

4. Polystyrene.

5. Stearic acid ($C_{17}H_{35}COOH$)

(*i*) The use of a high temperature stationary phase.

(*ii*) Derivatisation.

(*iii*) Pyrolysis gas chromatography.

Learning Objectives

After studying the material in Part 5, you should now be able to:

- understand the factors affecting the upper temperature limit of stationary phases and know what steps to take to extend the upper temperature limit of a phase;
- discuss the reasons for derivatisation and choose an appropriate procedure for use with a given sample;
- describe the pyrolysers in common use and discuss their relative merits.

6. Qualitative Analysis by Gas Chromatography

6.1. INTRODUCTION

Like any other chromatographic technique, gas chromatography is essentially a separation process. That it can be turned into an analytical technique which yields qualitative and quantitative results should be considered a bonus. Such results may not come easily and you may need to work hard for them, but the effort is well worthwhile: gas chromatography can analyse mixtures with which other techniques cannot cope.

In its simplest form, qualitative analysis is performed by means of retention data. The retention of a compound will be determined by the position of its distribution equilibrium between the stationary and mobile phases, ie the distribution ratio. Its retention data will therefore be characteristic of it, though not uniquely so: it is quite possible for several compounds to have the same retention time. Nonetheless, with some knowledge of the sample and what it is likely to contain, it is possible to identify the components present. This can be done by comparison of their retention data with those of standard compounds. However, a retention time is only a single piece of information, and with only a retention time to go on, it is not possible to rule out coincidence. You really need many more data on an unknown compound, all of which can be shown to be identical to the data for a standard compound, before you can say with certainty that you have identified it. In this respect, gas chromatography is bound to be far inferior to techniques such as infrared spectrometry. Infrared spectra are usually quite complex and often allow unequivocal identification because of the large number of bands at different wavenumbers. Furthermore, there is a close correlation between the wavenumbers of bands in an infrared spectrum and the presence of certain structural features in a compound. It is possible to go a long way towards elucidating the structure of a completely unknown compound by infrared spectrometry, whereas gas chromatography is much less informative.

On the other hand, a mixture of two or more compounds is likely to cause great difficulty in infrared spectrometry. The large number of bands, which in one respect we have seen to be an advantage, becomes a problem. With so many bands present, many of them are bound to overlap and there is no easy way of deciding which belong to each component. Interpretation of each spectrum will be very difficult if not impossible. The single peak given for each component of a mixture by gas chromatography is much less likely to overlap or interfere with others. Since it is only human to want the best of both worlds, many very able chemists and instrument engineers have worked hard to combine the separating power of gas chromatography with the structural information and certainty of identification of molecular spectrometry. This has not always been easy, since gas chromatography presents the separated components in a form that is unsuitable for most spectrometric techniques.

In the following sections of Part 6 we shall see how the two approaches to qualitative analysis by glc (the use of retention data alone and in combination with a secondary technique) work out in practice.

6.2. COMPARISON OF RETENTION DATA

As you are probably aware, identification in chromatography is achieved by the comparison of retention data. However, retention times vary with changing experimental conditions and this problem can be particularly serious in gas chromatography. In practice, the 'relative retention (r) of James and Martin and the related 'retention index (I)' of Kovats are the only satisfactory answers. These both compare the retention of each analyte with that of a reference compound or compounds, chromatographed under the same experimental conditions. The retention of each analyte is then measured relative to the reference(s), so that much more consistent results can be obtained. My own preference is for adding the reference compounds to the mixture before analysis, so that it is subjected to *exactly* the same conditions as the analyte. It is quite possible, however, to inject it separately on the assumption that conditions will not change significantly between analyses.

6.2.1. Relative Retention

The rate at which a component passes through a chromatographic column, and hence its retention time, is controlled by its 'distribution ratio'. We normally think of this as the ratio of the amounts of a component present in the stationary phase and mobile phase at any time. The equilibrium between the two phases is dynamic, so we could also think of it as the ratio of the times spent by component molecules in the stationary and mobile phases. Since it is the stationary phase which differentiates between components, it is the proportion of time spent in the stationary phase that is significant. Most useful comparisons are therefore made by subtracting the retention time (t_M) of a non-retained compound which spends none of its time in the stationary phase, eg the 'air peak', from the measured retention time (t_R). The result is the adjusted retention time (t'_R):

$$t'_R = t_R - t_M$$

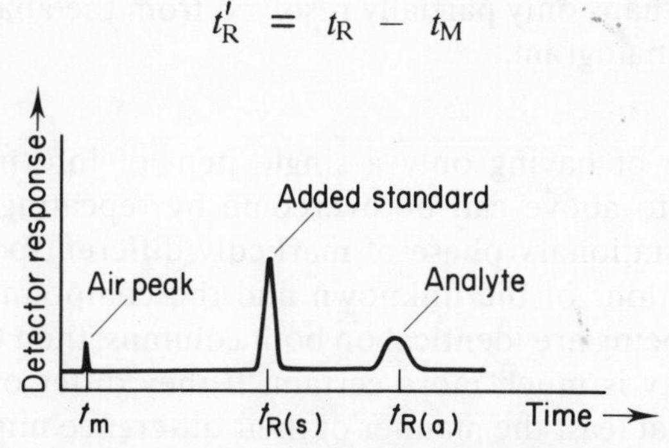

Fig. 6.2a. *Chromatogram showing an 'air peak', analyte (a), and added reference compound (s)*

The relative retention of an analyte (a), compared with an added reference compound (s), may be calculated as follows:

$$r_{(a,s)} = \frac{t'_{R(a)}}{t'_{R(s)}} = \frac{t_{R(a)} - t_M}{t_{R(s)} - t_M}$$

The relative retention of a compound measured in this way will vary only with the temperature and the stationary phase, and should otherwise be independent of other experimental conditions. This makes it fairly reproducible between laboratories. Provided the standard, the stationary phase and the temperature are quoted, the comparison of such relative retentions is a reasonable basis for identification. It is even possible to allow for differences in temperature, since a plot of the logarithm of the relative retention of a compound against the reciprocal of the absolute temperature is linear, but in my experience this is not often done. It is easier to work at the same temperature as the one used for the original data.

There is always some degree of error in measuring retention times, so that a relative retention based upon the ratio between a small retention time and a large one might not be very accurate. To overcome the uncertainty in identification which arises because of this, you can use the technique of 'spiking'. When you think that you have identified your analyte, you add a small amount of it to your sample (you 'spike' your sample). If your identification was correct, the analyte peak simply increases in size. If it was incorrect, then a new peak, perhaps only partially resolved from the analyte, appears on your chromatogram.

The difficulty of having only a single item of information which was referred to above can be overcome by repeating the analysis on a second stationary phase of markedly different polarity. If the relative retentions of the unknown and the compound which it is suspected of being are identical on both columns, then the presumption of identity is much more certain. If they differ on the second column, then at least the manner of their difference might give some further clue as to the polarity and characteristics of the unknown compound.

Π An unknown compound, J, has the following relative retentions compared with hexane:

Stationary phase	Temp. (°C)	$r_{J,hexane}$
Squalane	65	0.23
PEG-400	65	10.5
PEG-S	65	2.5

To which of the following classes is J most likely to belong?

(*a*) Alkane

(*b*) Alcohol

(*c*) Alkene

(*d*) Ketone

If you suggested (*b*), an alcohol, good. You have remembered the ideas that we discussed in Part 3.3 – the Stationary Phase. The relative retentions of less than one on squalane and more than one on PEG-S indicate that J is much more polar than hexane, and this rules out the alkene. The much higher relative retention on PEG-400 than on PEG-S suggests a high degree of hydrogen bonding and indicates an alcohol.

You could even suggest that 'J' was likely to have a boiling point in the same sort of region as hexane – say between 65 °C and 85 °C. This would give you some idea of which alcohols to compare the retention of J with. I would ask you to bear in mind that I have chosen a fairly simple example to illustrate the principle. In practice it is usually less straightforward than this!

Quite often the compound which you think your analyte might be, and with which you want to compare it, is not available and its relative retention is not quoted in the literature. In this case, it is sometimes possible to estimate what the relative retention would be by taking advantage of the extrapolation or interpolation of the relation between retention and certain other molecular parameters for a homologous series. Thus the plots of the logarithm of the relative retention against the number of carbon atoms or against the boiling point are both linear for any given homologous series. If the retention data for a few members of the series are available then it is possible to estimate the data for the remainder. Of course, you do not always know the homologous series to which your unknown compound belongs. The best that you can do then is to collect the data using a relatively non-selective stationary phase, such as a polydimethylsiloxane. On such a stationary phase the plots for boiling points for different homologous series tend to merge into a single straight line. From this it is at least possible to get an estimate of the boiling point of a completely unknown compound.

One of the most serious difficulties with the use of relative retentions is in the choice of the reference compound for addition to the mixture to be analysed. The relative retention is a ratio of two numbers. If these numbers are not within an order of magnitude of each other, then any errors in measurement become exaggerated during calculation of the ratio. The added reference compound has, therefore, to be chosen so that it elutes not too far from the component whose relative retention is being determined. If a mixture of compounds is being examined, then the added reference compound should elute near to the middle of the mixture. Needless to say, it should also be resolved from all the components of the mixture! These requirements do limit the choice of reference compounds and also lead to many different ones being used. This is not too bad for work within a single laboratory, where there will be a limited range of interests, but the multiplicity of data is a little confusing when it comes to publishing work or compiling a nationwide data bank. It was to solve this difficulty that the retention index was proposed.

6.2.2. Retention Indices

The most useful system of retention indices is the one due to Kovats. It takes advantage of the linear relation between the logarithms of the adjusted retention times of a homologous series (the *n*-alkanes) and the number of carbon atoms in the molecules.

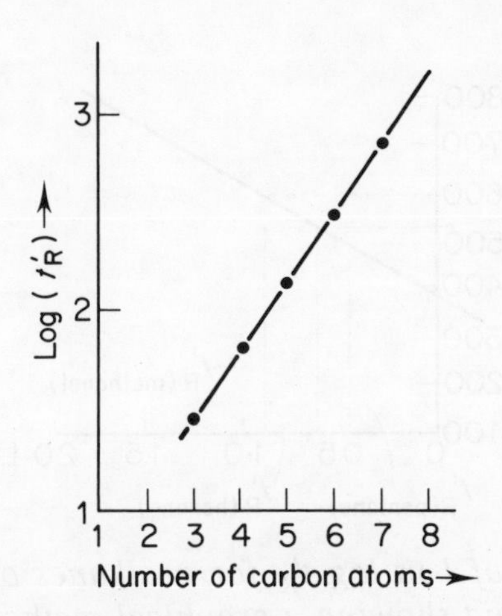

Fig. 6.2b. *Graph of log (t'_R) on DNP at 100 °C vs number of carbon atoms for the n-alkanes*

The *n*-alkanes are used as the reference compounds because of their stability, ready availability, cheapness and wide range of boiling points (which means that they can cover a wide range of analytes). The retention of any analyte is compared with the two *n*-alkanes which elute nearest to it. The adjusted retention time of the analyte is measured at the same time as those of the *n*-alkanes which elute in front and behind it (containing 'z' and '$z + 1$' carbon atoms respectively), and the retention index of the analyte, I, is then defined by:

$$I = 100 \times \left[\frac{\log t'_{R_{(\text{subst})}} - \log t'_{R_{(n\text{-C}_z)}}}{\log t'_{R_{(n\text{-C}_{z+2})}} - \log t'_{R_{(n\text{-C}_z)}}} + z \right]$$

For the *n*-alkanes, the term

$$(\log t'_{R_{(\text{subst})}} - \log t'_{R_{(n\text{-C}_z)}})$$

reduces to zero and they thus have retention indices equal to the number of carbon atoms in the molecule multiplied by one hundred.

A graph of retention index against the logarithms of the adjusted retention times for the alkanes would therefore be a straight line (Fig. 6.2c).

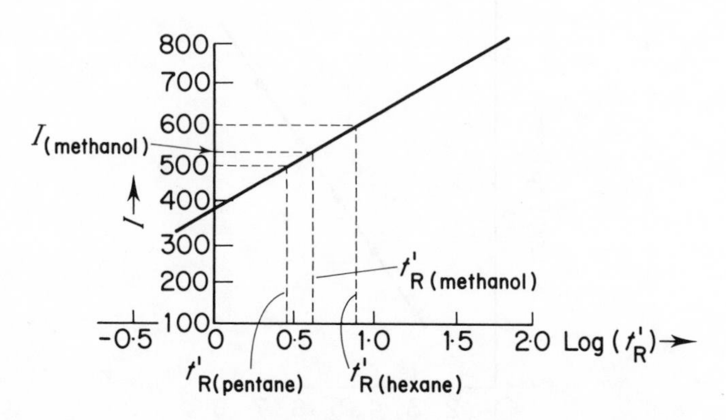

Fig. 6.2c. *Graph of I vs log t'_R for n-alkanes on DNP at 100 °C showing linearity and showing a graphical method for determining I for methanol*

Methanol, under the same conditions as those used for the data from which Fig. 6.2c was constructed, has an adjusted retention time of 4.50 min, eluting between *n*-pentane (3.05 min) and *n*-hexane (7.70 min).

Either

interpolation on the graph (see Fig. 6.2c)

or

calculation

$$I = 100 \times \frac{\log 4.50 - \log 3.05}{\log 7.70 - \log 3.05} + 5$$

should give a value for I of 542. In effect, then, the retention index of a compound is equal to the number of carbon atoms (usually fractional) in the hypothetical *n*-alkane which has the same adjusted retention time, multiplied by one hundred.

Retention indices may vary slightly with temperature and markedly with stationary phase, but they are otherwise quite independent of conditions. This makes them fairly reliable. The retention index concept is very powerful and there is a very extensive literature of published values. I would have to say, though, that the majority of the analysts in small to medium sized laboratories in local industry with whom I come into contact are more likely to be using simple relative retentions.

SAQ 6.2a

Which of the compounds in List A would you choose to use as an added standard compound in order to measure the relative retentions of the components of the mixture of chlorinated aromatic hydrocarbons whose chromatogram is shown below?

List A

Compound	t_R (min)
benzene	0.85
bromobenzene	3.9
2,4-dichloromethylbenzene	5.9
butyl benzoate	6.2

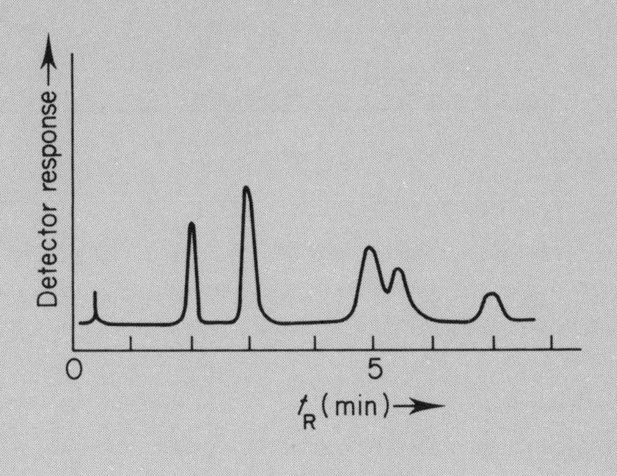

SAQ 6.2a

SAQ 6.2b

Calculate the Kovats Index of the compound eluting between decane and undecane on the chromatogram below.

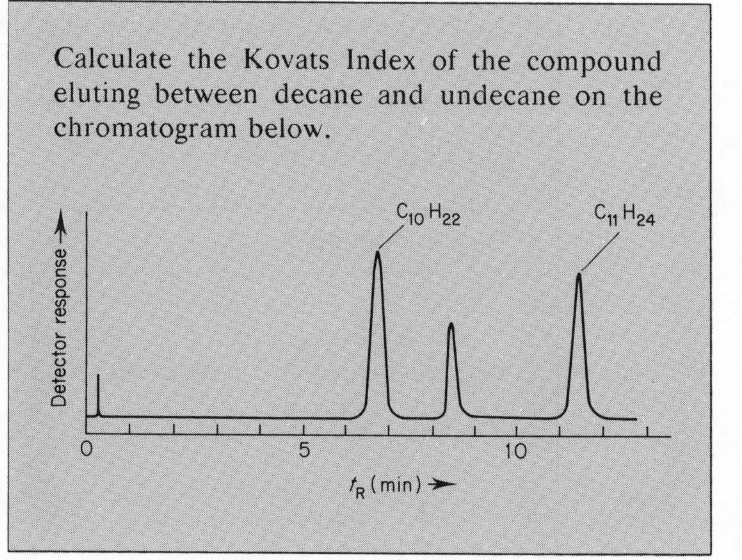

6.3. THE RECOVERY OF SAMPLES FOR SPECTROMETRIC ANALYSIS

Early attempts to improve qualitative analysis by gas chromatography have much in common with preparative gas chromatography since both are concerned with methods of recovering liquid samples from effluent gas streams. As we have seen, gc does not tell you very much about a completely unknown compound – a retention time, which is of very little use until you know what to compare it with, an estimate of a boiling point from a log (retention time) *vs* bp plot on a silicone column, some clues as to its polarity from the way its relative retention changes on different stationary phases, perhaps some idea of the elements present if selective detectors are used. Molecular spectrometry, in its various forms, is far more informative. Early workers saw that there was an obvious benefit to be gained by combining the two techniques. Unfortunately, there was also a disparity between the way in which the spectrometers then available needed to have samples presented to them and the condition in which separated components emerged from a gas chromatograph.

A gas chromatograph is at its most efficient when very small samples are used. The separated components then emerge as very small amounts mixed with a vary large amount of carrier gas in a fast moving stream at a little above atmospheric pressure.

Π The chromatogram below was obtained from a 1 μl sample of benzene injected onto a gas chromatograph using nitrogen as carrier gas, at a flow rate of 40 cm^3 min^{-1}, and an oven temperature of 42 °C.

(*a*) For approximately how long is there benzene in the carrier gas emerging from the column? (ie how 'wide' is the peak, in minutes?)

(*b*) Is the average concentration of benzene in the effluent gas stream:

(*i*) approximately 1% v/v?

(*ii*) approximately 10% v/v?

(*iii*) approximately 50% v/v?

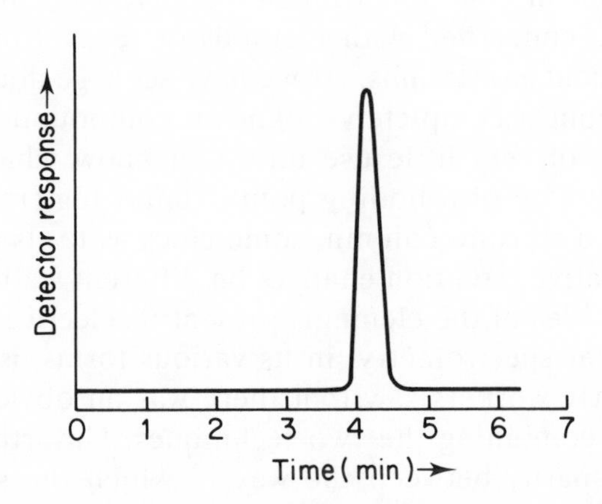

Fig. 6.3. *Chromatogram of benzene at 42 °C and a flow rate of 40 cm^3 min^{-1}*

(*a*) If you measured the 'width' of the peak, it took about 0.7 min to pass through the detector, ie *about 40 seconds*. (If, however, you only consider the time during which the concentration of benzene is more significant, say more than one third of its maximum value, this reduces to about 20 seconds).

(*b*) The correct answer is *approximately 1%*. I worked it out as follows:

Benzene has a density of about 0.8 g dm^{-3} and a RMM of 80.

1 μl of benzene will therefore weigh about 0.8 mg, and thus represents about 0.01 mmol.

During the time that the benzene emerged (0.7 min), nitrogen will have been flowing at 40 cm^3 min^{-1}, ie 0.7 × 40 cm^3, or approximately 30 cm^3, of nitrogen will also have emerged. The average concentration of the benzene will thus be:

$$0.01 \text{ mmol per } 30 \text{ cm}^3$$

Since a millimol of vapour occupies 22.4 cm^3, the v/v concentration of benzene will be:

$$\frac{0.224}{30} \times 100\% = 0.75\%$$

Given the assumptions and approximations made in the calculation, it is unrealistic to quote this any more precisely than 1% v/v.

At the time we are talking of (during the 1960's), a routine infrared spectrometer took about 15 minutes to record a full range spectrum and any attempt to speed this up lead to very poor spectra (loss of resolution, apparent shifts in band positions). Such an instrument would record a satisfactory spectrum of a liquid if about 50 μl were held between rock salt plates, or a vapour if about 50 cm^3 were placed in a 10 cm path length gas cell. Any attempt to record the spectrum from a 1 μl sample of liquid, vaporised and mixed with nitrogen so that its concentration increased from 0% to 1% and then decreased to 0%, all in 40 seconds, was out of the question.

Nuclear magnetic resonance (nmr) spectrometry required a similar size sample to ir spectrometry, but it had to be presented in solution in a rapidly spinning tube. Mass spectrometry (ms) required a much smaller sample, but it had to be admitted into the instrument without reducing the vacuum significantly. It would therefore be difficult, if not impossible, to cope with a sample in a carrier gas stream of 40 cm^3 min^{-1}!

We shall see later that many of these problems have now been overcome, but at the time it seemed more profitable to explore the recovery of the sample from the carrier gas as a liquid. It could then be presented to a spectrometer in a form suitable for recording a spectrum – the so called 'trap and transfer' technique. This would have a lot in common with preparative gas chromatography, and it is appropriate to discuss this technique before proceeding to directly linked gc/spectrometric techniques: there may be occasions when you will need to use the 'trap and transfer' technique (eg if you need the nmr spectrum of a component).

If gas chromatography is to be used for preparative purposes, there are two problems to be overcome. The first is to recover the separated components from the carrier gas and the second is to increase the amount of sample that can be handled.

6.3.1. Sample Trapping and Preparative Gas Chromatography

A common arrangement for this purpose is for a 'stream splitter' to be attached to the outlet end of the column. A small proportion (often about 1%) of the gas flow is allowed to pass to the detector whilst the remainder (99%) is directed to one of a number of traps. These are designed to retain the separated components whilst allowing the unwanted carrier gas to flow to waste. There will be a separate trap for each component.

The design of the stream splitter does not present serious problems. It follows along the same lines as the injection splitter discussed at the beginning of this Unit. Since it is necessary to avoid premature condensation of the components, the splitter is usually situated in the column oven and the 'recovery' stream is conveyed to the traps along a heated capillary tube.

The design of the traps is much more of a problem. Trapping efficiency is all important, and high efficiencies are not easy to achieve. The obvious approach is to condense the component in a cold trap. Such traps usually have a large surface area and a large volume, so that the sample has time to condense. Simple U-tubes, coils and combinations of bulbs and coils surrounded by a coolant (ice, solid carbon dioxide, liquid nitrogen), have all been used. Whatever the design of the trap, the most serious losses occur as a result of mist formation, and a small plume of 'smoke' coming from the trap outlet is a not uncommon sight. In general, if a hot gas stream containing a condensable component is cooled rapidly, the component molecules agglomerate to form a stable aerosol rather than liquify on the walls of the containing vessel. Using a coolant of only moderately low temperature, so that cooling is less rapid, is a partial answer to this problem, as is the careful design of the trap. Neither, however, can be completely successful.

Much more effective is the use of a 'packed trap'. This can take the form of a cooled U-tube filled with an inert powder, but filling it with a conventional glc packing is much more effective. The component is efficiently retained in the stationary phase, even at the very high carrier gas flows involved, and mist formation is not a problem. The component can then be expelled by warming the packed trap whilst passing a low flow of carrier gas through it, and recovered in a cold trap. At this low flow rate, mist formation in the cold trap is less of a problem and the overall trapping efficiency can be quite high.

6.3.2. Scaling Up

In analytical gc, a typical 4 mm id packed column, with a 10% stationary phase loading, is operating well within the limits of its capacity. The sample size can be increased quite significantly before there is a noticeable loss of resolution. If the components that are to be recovered are *very* well resolved to begin with, the sample size can be increased still more without the deterioration in resolution becoming prohibitive. Even so, there will be a limit to the size of the sample which can be separated by such a column. To some extent, the time it takes to inject *and vaporise* a very large sample will influence this. If it takes a long time, the resulting peaks will be broad. The time taken to vaporise the sample in an empty, heated zone at the front of the column is often the limiting factor and direct, on column injection is often more effective. The trouble with this is that repeated injections of large amounts of liquid onto the packing may strip the stationary phase from the support and so cause column deterioration. All things taken into account, it is doubtful if you should consider separating much more than 100 μl of liquid sample on a conventional analytical column. This can often be enough to recover a sample of a component sufficient for spectrometric analysis.

If you need to handle more sample than this, you have two alternatives – either to repeat the injection/separation/collection cycle several times with a smallish sample on an analytical column and then to combine the appropriate fractions from each cycle or to increase the capacity of the column. (In practice, since pushing any technique to extremes can lead to difficulties, you may well decide to compromise – to increase the capacity of the column a bit and repeat the collection cycle just a few times).

Repeating the cycle manually more than just a few times is going to be time consuming, tedious and therefore prone to error. Some degree of automation is essential, and suitably modified instruments have been manufactured for this purpose. They have usually taken the form of a conventional gas chromatograph fitted with an automatic injection device, a preparative column (see later), a splitter, a selection of traps connected to a switching device and a sequence controller. The keys to its successful operation are stability and reproducibility. The main functions of the sequence controller are to initiate a cycle by injection and then to activate the trap switching device at appropriate times so that the equivalent fractions from each successive cycle are collected in the same traps. If the chromatographic conditions are unstable or the sequence controller fails, all subsequent fractions will be collected in the wrong traps. Nonetheless, such instruments can be quite effective and much valuable work has been done with them.

One way of avoiding the problems of sequence controlling is simply to scale up the column dimensions so that much larger samples can be injected. Columns of up to 10 mm id can be accommodated in the oven of a conventional gas chromatograph, and this 65-fold increase in cross sectional area taken in conjunction with a doubling of the stationary phase loading to 20% means that the sample size can be increased by about one order of magnitude. However, in order to get a high enough carrier gas flow rate, larger particle diameter packings need to be used, with a consequent loss in performance.

It is also asking rather a lot of the injection heaters of an analytical instrument to provide enough heat to vaporise large samples rapidly. Nonetheless, respectable amounts of volatile mixtures can be separated in this way.

Π Why does using larger particle diameter packings lead to a loss of performance?

(*i*)

(*ii*)

Your answer should have included reference to:

(*i*) less even packing due to larger, more irregularly shaped particles and probably a wider particle size range.

(*ii*) smaller surface area to mass ratio, leading to thicker stationary phase films and larger void spaces between particles. Hence there will be slower mass transfer between phases and greater non-equilibrium.

If it did – well done. You have remembered this much from Section 3.4 of this Unit. If you were uncertain, perhaps a quick re-read of Section 3.4 is in order.

Wider columns than this (up to 10 cm id) are available for laboratory use, but of course they cannot simply be fitted into a conventional gas chromatograph, but must be built into a specially designed preparative chromatograph. Large samples, perhaps up to 200 cm^3, can be separated on them.

Increasing the diameter of the column brings with it a number of problems. The packing material appears to segregate so that the mean particle diameter is larger close to the column walls. The result of this is that the carrier gas can flow more rapidly through the larger spaces between the particles near the wall and serious band broadening follows. This broadening is made worse by the uneven flow patterns that result from using packings with larger particles and wider particle size ranges. On top of this, there is a heat transfer problem with such wide columns, and the poor thermostatting which can result from this leads to poorer separations. To give you some idea of the scale of the problem, it is not uncommon for the HETP for a 10 cm id preparative column to be about ten times the HETP for a 2 mm id analytical column. This represents a dramatic loss of separating power, even before the effects of overloading and the slow vaporisation of large samples are taken into account.

I have spent rather a long time discussing the difficulties of preparative gas chromatography; this is necessary if you are going to avoid all the pitfalls for the unwary. I hope, though, that it has not given you the impression that preparative gas chromatography is an unusable technique, because this is far from the truth. With a little practice and a realistic view of what is possible, it is a very valuable tool for the gas chromatographer.

Zlatkis and Pretorius have edited an exhaustive text on the subject, which you ought to read if you are considering using this technique.

SAQ 6.3a

List three differences between analytical and preparative columns for gc.

1.

2.

3.

SAQ 6.3b

(*i*) Which of the following traps would it be best to use for recovering diethyl phthalate (bp = 298 °C) in preparative gc?

1. a U-tube cooled in ice

2. a U-tube cooled in propanone/solid CO_2

(*ii*) Suggest a trap which would be more efficient than either of the above.

SAQ 6.3b

6.4. DIRECTLY LINKED GC/IR

Given the proven value of infrared spectrometry in organic qualitative analysis, it is an obvious candidate for use with gas chromatography. Preparative gas chromatography provides large enough samples for a trappped and condensed component to be physically transferred to a microcell and the spectrum recorded with a conventional spectrometer. But there is an obvious advantage in being able to use an analytical gas chromatograph linked directly to a spectrometer. The incompatibility between sample size requirements has become less through the use of micro-cells and beam condensers for liquid samples and multipath cells or light pipes for gas samples . However, scanning speeds of conventional spectrometers (1 to 10 mins) are still limited by the response time of the detectors used, and are in general far too slow for continuous monitoring of the gc effluent.

This problem has now been largely solved by modern Fourier transform ir spectrometers with their high 'scanning' speeds, but the techniques worked out by early investigators are still of value to analysts not yet equipped with one of these instruments. Usually, these techniques involved fitting an outlet splitter onto the column and using the diverted part of the carrier gas stream to obtain the spectrum. One approach to solving the problem posed by a sample whose concentration varied too rapidly for the spectrometer to record a meaningful spectrum was to 'freeze' it whilst scanning was taking place, by a technique known as 'peak parking'. This was done by passing the diverted column effluent through a multipath gas cell and stopping the carrier gas flow in the gas chromatograph when a component had entered the cell. The spectrum of the entrapped vapour was recorded and the carrier gas flow restarted as soon as the recording was complete. The interruption to the carrier gas flow, and the diffusion which occurred during the periods of 1 to 3 minutes when the flow was stopped, lead to some peak broadening and a slight loss of resolution.

Many variations on trapping were devised. The diverted carrier gas was bubbled through a small amount of a solvent (tetrachloromethane and carbon disulphide are valuable) when a component was emerging and the spectrum of the solution subsequently recorded. Alternatively, the gas stream was passed into a cold trap packed with powdered potassium bromide. The powder, with the entrapped component, was then transferred to a die for compression into a disc and the spectrum duly recorded. Again, a range of cold traps which could also act as micro-cells was designed so that transfer losses could be minimised. In general, though, such techniques are rarely used today.

The recent development of Fourier transform infrared spectrometers has allowed the very rapid acquisition of spectral data which can be transformed into conventional spectra and stored for subsequent computer analysis. The small volume of a light pipe (approximately 1 cm^3) compared with the larger volume of the earlier multipath cells (approximately 40 cm^3), permits the spectrum of a component passing through it to be recorded without stopping the carrier gas flow, within the time that the concentration is reasonably constant. It is even possible to record spectra at the front and rear of a peak and thereby determine whether it consists of two partly resolved components.

Π Nuclear magnetic resonance is just as useful as infrared spectrometry, and the sample sizes and scan times are not greatly different. Why, then, do you think we have not developed directly linked gc/nmr systems?

Did you remember my earlier comment about the samples for nmr needing to be spun very rapidly? It would be almost impossible, using present technology, to design a physical interface between gas chromatograph which could condense the separated components into a rapidly spinning tube in a nuclear magnetic resonance spectrometer, without leaks, and interference with the smooth spinning of the tube poses another problem.

Yes, nmr would be valuable for identifying components separated by gc. For the time being, though, we are limited to the 'trap and transfer' technique.

6.5. DIRECTLY LINKED GC/MS

Incompatibility with gas chromatography was initially much less of a problem for mass spectrometry than for other spectrometric techniques. The sample sizes required are quite similar and scan speeds are high. The greatest problem to be overcome is how to introduce the separated components into the mass spectrometer without the large amount of carrier gas that is mixed with them causing an excessive increase in pressure.

The simplest solution would be to attach an outlet splitter to the column and pass the diverted gas stream to the mass spectrometer inlet system by way of a suitable flow restrictor. The restrictor would be adjusted to reduce the gas flow into the spectrometer to a level at which its pumping system could cope. Only a small proportion of the separated component would thus find its way into the spectrometer, so the sensitivity would not be great. Because of this, the approach was not greatly used to begin with. However, with the widespread use of capillary columns with their very low carrier gas flow rates, and with the improvement in both the sensitivity and pumping systems of modern mass spectrometers, the entire carrier gas flow can now be passed to the spectrometer. There were, though, and indeed there still are, situations where such a simple interface is not adequate. It is then necessary to use some device that will increase the component/carrier gas ratio before admitting the diverted carrier gas flow into the mass spectrometer. Such devices, called *molecular separators* or just *separators*, usually rely upon differences in rates of diffusion for the component and the carrier gas, which is usually helium. Separated components, because of their higher relative molecular masses, diffuse much more slowly.

Watson and Bieman designed a separator which consisted of a very fine porosity sintered glass tube about 10 cm long, enclosed within an outer tube of glass (see Fig. 6.5a). The diverted gas flow enters the porous glass tube through a flow restriction and the 'concentrated' gas stream is drawn into the mass spectrometer through a restriction at the other end.

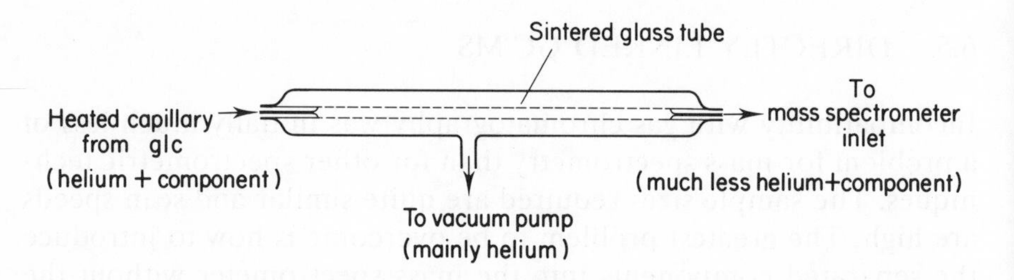

Fig. 6.5a. *A simple Watson–Boeman separator (Other geometries are possible and the sintered glass tube can be replaced by ceramic or metal frits)*

Helium diffuses through the porous glass more rapidly than the component and is pumped off, leaving a more concentrated gas mixture within the tube. Depending on the RMM of the component, the porosity of the frit and the pumping rates employed, the concentration of the component can be increased by a factor of anything between two and ten. However, to achieve a greater increase in concentration the rate at which the waste gases are pumped off has to be so high that much of the component diffuses through the frit as well, and the amount entering the spectrometer is quite low. Some compromise usually gives better overall results.

Ryhage designed a separator which relied upon the different rates of diffusion of gases in an expanding supersonic jet stream (See Fig. 6.5b).

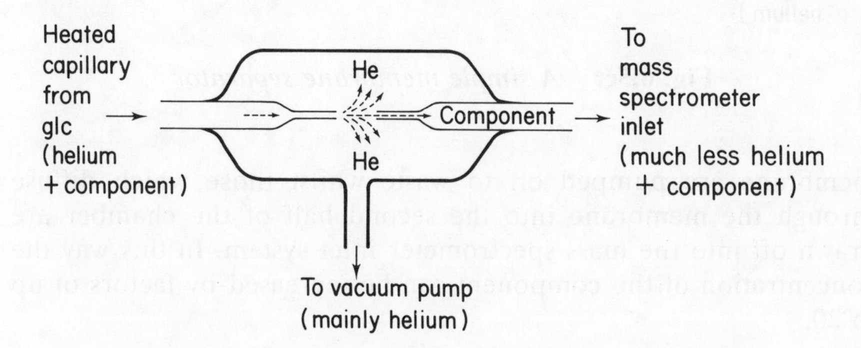

Fig. 6.5b. *A schematic diagram of a jet separator*

If the diverted gas stream is allowed to expand into an evacuated chamber, the helium diffuses more rapidly in a transverse direction, away from the jet due to its lower RMM. If a second jet is placed in line with the first, it can be used to draw off the residual, axial stream of gas, which now has an increased concentration of the separated component.

An alternative approach has been to take advantage of the solubility of organic compounds in silicones. This allows them to diffuse through very thin silicone rubber membranes much more rapidly than helium, so that it is possible partially to separate organic components in the diverted gas stream from the helium used as carrier gas. Several separators have been designed (See Fig. 6.5c).

In general the diverted carrier gas stream is passed into a chamber divided in two by a silicone rubber membrane supported on a glass sinter. Gases which do not diffuse through the

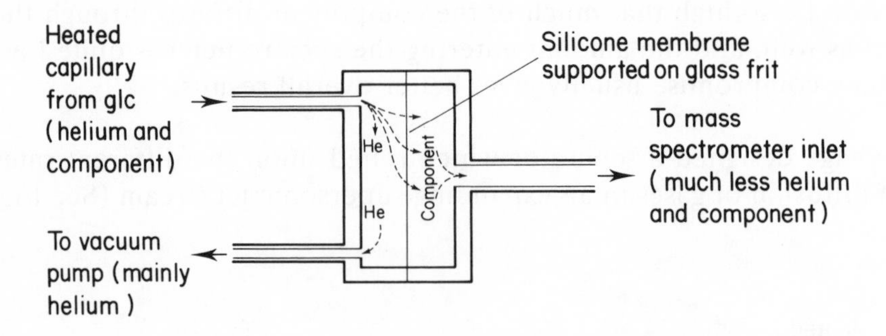

Fig. 6.5c. *A simple membrane separator*

membrane are pumped off to waste whilst those which diffuse through the membrane into the second half of the chamber are drawn off into the mass spectrometer inlet system. In this way the concentration of the component can be increased by factors of up to 20.

Each of these separators will perform slightly better under different circumstances and their relative merits have been discussed fully by Mcfadden. I would certainly recommend that you consult this book if you wish to study gc/ms in more detail.

The techniques for separating quite complex mixtures by gc and obtaining quite respectable mass spectra of the separated components have been available for some time now and represent a considerable advance in qualitative analysis of organic mixtures, even if it is rather an expensive one. The amount of data that is acquired during such an analysis is enormous. It would be very inefficient to rely upon manual recording and filing of such data and almost impossible to interpret it all manually. The advances made in the use of gc/ms are as much due to the use of dedicated computers which record and store the data from an analysis as to the development of efficient separators. Such computers can be programmed to undertake searches of extensive libraries of spectra, make identifications and present full analyses of mixtures. The combination is enormously powerful and the extensive applications that have been reported bear witness to this.

Summary

Gas chromatography is quite commonly used for qualitative analysis by comparing the retention data of the analyte with those of the compound which it is thought to be. Simple retention times are not very reproducible and it is better to use relative retentions or retention indices.

If you have no idea what the analyte might be, you will need to combine gas chromatography with spectrometry so that interpretation of the spectrum of the analyte can help in its identification. This may be done by the preparative separation of the components of a mixture by gas chromatography followed by conventional spectrometry or it may be done by linking a gas chromatograph to the spectrometer for direct on line work. The latter approach has the benefit of speed and sensitivity, but is far more expensive. It has been particularly successful in the case of mass spectrometry when used in conjunction with computer storage and processing of data. It has also been successful in the case of infrared spectrometry, particularly with the development of a new generation of rapid scanning Fourier transform instruments, but for most other spectrometric techniques, 'trap and transfer' is the order of the day.

SAQ 6.5a

Which of the following techniques can be said to be amenable to being usefully and effectively 'directly-linked' with gas chromatography so that spectra can be recorded without interrupting the chromatographic process (recorded 'on the fly')?

(*i*) infrared spectrometry

(*ii*) ultraviolet spectrometry

(*iii*) nuclear magnetic resonance spectrometry

(*iv*) mass spectrometry.

SAQ 6.5a

SAQ 6.5b

Select from the following list those reasons for linking a gas chromatograph and a mass spectrometer which you think are valid and place them in a rank order (the most valid first).

(*i*) to reduce costs

(*ii*) to check whether a peak on a gas chromatogram represents one component or two unseparated components

(*iii*) to improve quantitative analysis

(*iv*) to identify unknown peaks

(*v*) to improve detector sensitivity.

SAQ 6.5b

Learning Objectives

After studying the material in Part 6, you should now be able to:

- describe identification by retention data and recognise its limitations;
- calculate a relative retention and a Kovats index from a chromatogram;
- discuss the difficulties of on-line gc/molecular spectrometric techniques;
- describe techniques for on-line gc/ir and gc/ms;
- describe techniques for preparative gas chromatography.

7. Quantitative Analysis by Gas Chromatography

7.1. INTRODUCTION

The quantitative analysis of volatile mixtures containing components whose identity is already known probably constitutes the biggest single application of gas chromatography. We now need to consider how such analyses are performed. I shall assume, as I did in the treatment of qualitative analysis, that, if necessary, you have read and assimilated the ACOL Unit: *Chromatographic Separations* where the general problems of quantitative analysis by chromatography were considered. We shall not be looking in detail at the various ways of measuring peak areas, the problems posed by partially unresolved peaks or the advantages of integrators. I shall be concerned here only with those aspects that are relevant to gas chromatography.

Π Indicate, by circling either T for True or F for False, whether you agree with the following:

The areas of symmetrical, well resolved peaks can be measured satisfactorily by:

(*a*) peak height × width at half height

T / F

(*b*) an electronic (computing) integrator

T / F

(*c*) peak height × retention time T / F

(*d*) cutting them out and weighing them T / F

(*e*) measurement of the perimeter of the peak T / F

Now for each method for which you have answered, T, select a response from (*i*) to (*iv*) which represents the precision which you believe is associated with the method:

(*i*) < 10%
(*ii*) < 5%
(*iii*) < 2.5%
(*iv*) < 1%

My answers would be as follows:

(*a*) T < 2.5%

(*b*) T < 1%

(*c*) F

(*d*) T < 2.5%

(*e*) F

(*c*), peak height × retention time, does not measure the peak area. On the other hand, since peak widths are related to retention times and are susceptible to much more accurate measurement, some analysts have used this as a measure of peak area for preparing a calibration graph. I am not very happy with it, but don't ask me why! Pure prejudice, I expect, because it is not *actually* measuring the peak area. The parameter correlates quite well with the quantity of analyte present; you may like to try preparing a calibration graph and comparing it with the results from an integrator.

(*e*), peak perimeter, is False (think about a short fat peak and a tall thin peak of similar area!) You might have been thinking of a planimeter which traces around the outline of a peak, but measures the area (after a fashion) not the perimeter. My own view is that the least said about planimeters the better. They are alright for chunky areas (areas of towns on maps) but not for tall thin areas like chromatographic peaks.

The estimates of precision for (*a*) and (*d*) are based upon the results obtainable if you are very careful and choose a chart speed that gives wide peaks so that errors on width measurement are minimised. They are not easy to achieve. It is a lot easier to get good results with an electronic (computing) integrator.

If you do not feel confident after trying this question, it would be a good idea to read or to re-read the relevent sections of ACOL: *Chromatographic Separations*.

On the subject of integrators, it is sufficient to say that the much heralded *microprocessor revolution* has now reduced the price of quite sophisticated integrators to the point where it is more cost effective to invest in one than to use valuable technician's time on manual integration. It is only if you are involved in a small scale operation that you will not be able to justify buying an integrator and will be reduced to manual integration. Under these circumstances, my own preference is to use *peak height* × *width at half height* with symmetrical, well resolved peaks. I use a reasonably high chart speed and a magnifier with a scale graduated to 0.1 mm to reduce the error on width measurement, which is the greatest source of error. If, on the other hand, the peaks are badly tailed or only partly resolved (ie have a resolution, R_S, of less than 1.) I prefer to use cutting out and weighing, which I find more accurate.

Π Draw the dividing line which you would use in distinguishing, for manual integration, the areas of each of the overlapping peaks given below:

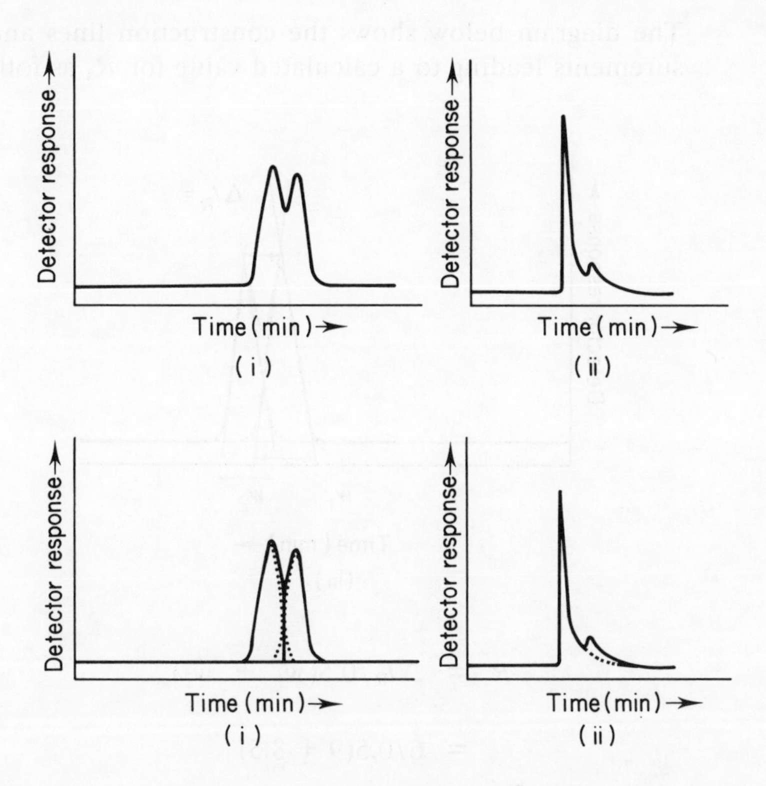

In (*i*), the two symmetrical peaks have a resolution of less than 1. You might like to practice working R_S out (see supplementary answer if you do). Each contributes some of its area to the other (broken lines), so that in dropping a perpendicular from the valley bottom we are in effect assuming that they cancel each other out. This works well if they are similar sizes, but even if one is about ten times as large as the other, the error is likely to be only about 10%.

In (*ii*) the small peak is sitting on the tail of the large peak and there is not the same *exchange* of areas. You have to sketch in the tail of the larger peak (called *peak skimming*).

Supplementary Answer:

My result for R_S was 0.7; R_S being the difference in retention times between two peaks divided by the mean peak width.

The diagram below shows the construction lines and measurements leading to a calculated value for R_S as follows:

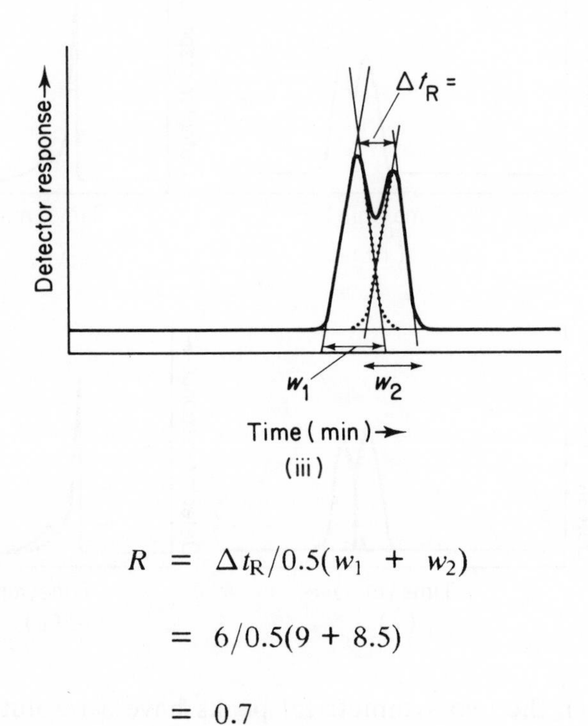

(iii)

$$R = \Delta t_R / 0.5(w_1 + w_2)$$

$$= 6/0.5(9 + 8.5)$$

$$= 0.7$$

If you were uncertain of the answers to this question, you ought to read or re-read the relevant section of the ACOL: *Chromatographic Separations*, before proceeding.

7.2. INJECTION AND QUANTITATION

The difficulty of accurately injecting small quantities of liquids must be the greatest limitation on quantitative gas chromatography. The use of a syringe or microsyringe in conjunction with a conventional injection port is prone to many errors. The use of an injection splitter with capillary columns is prone to even more. As a result of this, it is not realistic to contemplate the preparation of a single calibration graph by injecting a series of standards of different concentration (or different sized samples of the same standard) and then using it to determine the concentration of an unknown sample. It is essential that there be some means, internal to the sample, which allows for the variation in the size of sample and the effectiveness and speed with which it is applied to the column.

In gas chromatography, *area normalisation* and *internal standards* are used for this purpose. Area normalisation, in its simplest form, assumes that the sensitivity of the detector is the same for each component in the mixture. A more satisfactory version corrects for any significant variation in sensitivity. The composition of the mixture is then obtained by expressing the area of each individual peak as a percentage of the total area of all the peaks in the chromatogram. The assumption upon which area normalisation is based is reasonable if a narrow range of components of similar volatility and molecular structure is being analysed. It is much less reasonable in the presence of a variable amount of a component that is not detected (eg water when using a FID, or a non-volatile component with any detector). Neither is it reasonable when the components are *very* different, for any differences in thermal conductivity or carbon content will lead to differences in the sensitivity of TCD or FID, respectively. Under these circumstances, it is necessary to modify the simple area normalisation procedure by introducing relative response factors. This can be done by preparing and analysing standard mixtures of the analytes. The ratio of peak area/weight of component is then measured for each component and expressed relative to that of one of the components. The result is the relative response factor for each component. In subsequent analyses, multiplying each peak area by the appropriate relative response factor before normalisation gives a satisfactory analysis.

Area normalisation is quite commonly used for analyses that are performed repeatedly on very similar volatile samples – say for monitoring the composition of a solvent mixture in the paint industry or the impurities in a volatile product in the manufacturing side of the chemical industry.

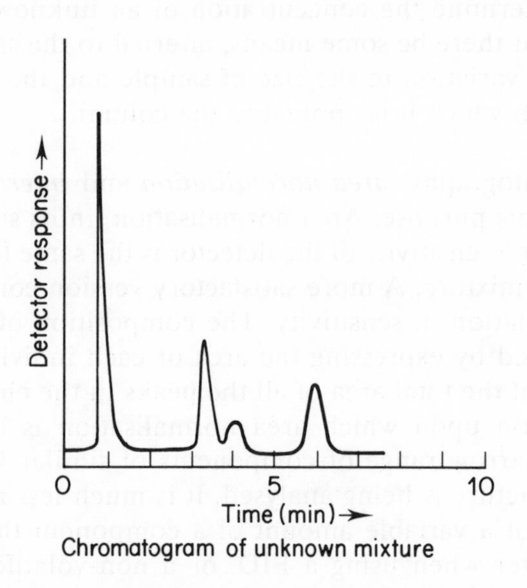

Fig. 7.2a. *Chromatogram of unknown mixture*

Π The above chromatogram was obtained when a mixture of unknown origin was analysed, using a gas chromatograph fitted with a katharometer (TCD), an integrator and a squalane column, and operating at 60 °C. The integrator readings were as follows:

Peak	t_R (min)	Integrator counts
1	0.8	24470
2	3.3	11171
3	4.1	4160
4	6.0	10816

Calculate an approximate composition for the mixture from the integrator readings, using simple area normalisation.

Now measure the peak areas manually (either by *peak height × width at half height* or by cutting out and weighing peaks using a photocopy) and recalculate the composition.

The components 1 to 4 were subsequently identified as ethanol, benzene, hexane and methylbenzene, respectively. A mixture of equal weights of these four components was consequently injected, with the following results:

Peak	Integrator counts
1	15600
2	12674
3	14230
4	12410

Calculate a relative response factor (relative to benzene = 1) for each of them and use these to calculate a more accurate composition of the mixture, using full, compensated area normalisation.

Compare this with your earlier results.

How did you get on? My results were as follows:

	Normalisation method		
Peak	Simple		Compensated
	Integrator	Manual	Integrator
1	48.3%	44%	43.4%
2	22.1%	24%	24.4%
3	8.2%	10%	8.1%
4	21.4%	22%	24.1%

If you take compensated area normalisation using an integrator as the most satisfactory result, you can see how large the errors can be if simple normalisation or manual integration are used.

If you had difficulty, you might like to follow my working through:

(*i*) Simple normalisation-integrator

Peak	Integrator	% Composition
1	24470	$\frac{24470}{50617} \times 100\% = 48.3\%$
2	11171	$\frac{11171}{50617} \times 100\% = 22.2\%$
3	4160	$\frac{4160}{50617} \times 100\% = 8.2\%$
4	10816	$\frac{10816}{50617} \times 100\% = 21.4\%$
Total	50617	– 100%

(*ii*) Simple normalisation-manual

Peak	Pk. ht. (mm)	Width at hlf. ht. (mm)	Area (mm^2)	% Composition
1	44	1	$44 \times 1 = 44$	$\frac{44 \times 100}{100.5} = 44\%$
2	15	1.6	$15 \times 1.6 = 24$	$\frac{24 \times 100}{100.5} = 24\%$
3	4	2.5	$4 \times 2.5 = 10$	$\frac{10 \times 100}{100.5} = 10\%$
4	9	2.5	$9 \times 2.5 = 22.5$	$\frac{22.5 \times 100}{100.5} = 22\%$
			TOTAL $= 100.5$	$= 100\%$

(*iii*) Compensated normalisation-integrator

from the standard mixture:

Peak	Integrator	Response factor
1	15600	15600/12674 = 1.23
2	12674	12674/12674 = 1.00
3	14230	14230/12674 = 1.12
4	12410	12410/12674 = 0.98

then, for the sample:

Peak	Integrator	Compensated area	% Compensation
1	24470	24470/1.23 = 19894	$\frac{19894}{45816} \times 100\% = 43.4\%$
2	11171	11171/1.00 = 11171	$\frac{11171}{45816} \times 100\% = 24.4\%$
3	4160	4160/1.12 = 3714	$\frac{3714}{45816} \times 100\% = 8.1\%$
4	10816	10816/0.98 = 11037	$\frac{11037}{45816} \times 100\% = 24.1\%$
		Total = 45816	= 100%

In principle, the use of an internal standard is more satisfactory than normalisation. It overcomes both the difficulties caused by variable amounts of a component that is not detected and variable evaporation losses from the syringe needle. It will also be much more satisfactory where a preliminary clean up and derivatisation is involved, since it will allow for any losses during those processes. It does, however, involve much more effort. First an internal standard must be chosen and then the calibration graph must be produced. However, manual quantitation can often be simplified when using an internal standard because the ratio of peak heights can be used instead of the ratio of peak areas. This is possible because any variation in conditions affects both analyte and internal standard alike and the ratios of peak heights to peak areas remain unchanged. Using peak height ratios in this way can sometimes reduce errors during manual integration. It eliminates measuring the width of narrow peaks which is a major source of error. The choice of internal standard is crucial. It must be as similar to the analyte(s) as possible. This will ensure that it suffers similar losses during clean up, derivatisation and injection and also that variations in chromatographic conditions and detector parameters affect both internal standard and analyte(s) similarly. The internal standard must also be resolved from all the components present, and should, ideally, be eluted somewhere near the middle of the mixture.

Once a suitable internal standard has been chosen, this is probably the most reliable method available for quantitative analysis. It is particularly convenient when only one or two components are of interest and when the analysis of trace components is involved.

Π The concentration of ethanol in a blood sample has been determined by the method of internal standards. A series of standard aqueous solutions of ethanol and the blood sample were all treated alike. An aliquot portion (0.01 μl) of each was diluted with a standard aqueous solution of propan-2-ol (0.1 μl; concentration = 0.1 mg cm^3). Samples (1 μl) of each dilution were then injected into a gas chromatograph fitted with an FID, an integrator and a PEG-400 column at 80 °C. The following results were obtained:

Sample	Integrator counts	
	Ethanol peak	Propanol peak
ethanol standards:		
0.5 mg cm^{-3}	5518	12754
0.75 mg cm^{-3}	7563	11893
1.0 mg cm^{-3}	10350	12084
1.25 mg cm^{-3}	13935	12870
1.5 mg cm^{-3}	15628	12314
blood sample:	9862	12604

Plot an appropriate graph and use it to determine the concentration of ethanol in the blood sample.

I made it 0.91 mg cm^{-3}. You should have got something between 0.88 and 0.94 mg cm^{-3}. I worked it out something like this: firstly, I extended the table of results to include the ratio of the areas of the ethanol and propan-2-ol peaks, then I plotted a graph of these ratios against the concentrations of the standard ethanol solutions, and finally, I read the result for the blood sample from the graph. It was possible to work with the concentrations of the original ethanol solutions, not making a calculation of the concentrations after dilution with the propan-2-ol solution, because standards and sample were treated exactly alike.

Sample	Ethanol peak	Propanol peak	Integrator counts Ethanol/Propanol
ethanol standards:			
0.5 mg cm^{-3}	5518	12754	5518/12754 = 0.433
0.75 mg cm^{-3}	7563	11893	7563/11893 = 0.634
1.0 mg cm^{-3}	10350	12084	10350/12084 = 0.857
1.25 mg cm^{-3}	13935	12870	13935/12870 = 1.08
1.5 mg cm^{-3}	15628	12314	15628/12314 = 1.27
blood sample:	9862	12604	9862/12604 = 0.782

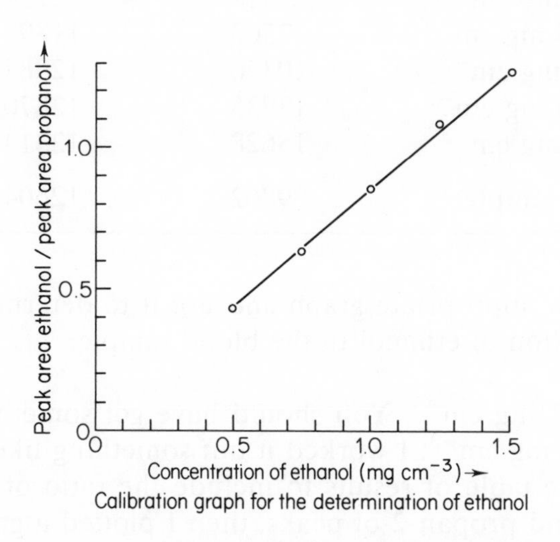

Fig. 7.2b. *Calibration graph for the determination of ethanol*

From the graph, the concentration of ethanol in the blood sample was 0.91 mg cm^{-3}.

7.3. GAS ANALYSIS

Gas analysis by gas chromatography presents its own problems. Although the quantitative addition of a suitable internal standard to a gas sample may be possible, it is far from simple and the method of internal standards is not used. Area normalisation is possible, but it is rather unsatisfactory because the TCD (katharometer), which is the detector most commonly used for gas analysis, varies in its sensitivity for different gases. The only alternative is simple calibration by the injection of different volumes of standard gas mixtures. We saw, in Section 2.5 of this Unit, that syringes are inherently unreliable when used for injecting gases and that gas sampling valves were developed to get around this difficulty. It is the accuracy of these valves which makes simple calibration the method of choice for gas analysis. As well as gas sampling valves, simple calibration relies upon the availability of accurate standard calibration mixtures of the gases to be analysed. Many of these are commercially available, at a price. If, however, you wish to analyse unusual gases you may need to contemplate using a vacuum rack to make up your own standards. You will probably also need to consider how to collect and transport your gas samples. The use of glass and stainless steel gas pipettes (elongated vessels with a tap at each end so that they can be flushed through with the gas and then closed off) is the most satisfactory solution, but in an emergency I have used toy balloons. It looks a little silly and gases will diffuse both into and out of the balloon through the thin rubber membrane, but reasonable results can be obtained if there is not too long a delay. I have even used a beer bottle to sample the atmosphere in a pub cellar where there was a strong smell of hydrocarbon gases! The trick is to wash it well, fill it to the brim with cold water, take it into the cellar, pour all the water out and then stopper it. Back at the laboratory, a sample can be withdrawn with a syringe and transferred to the gas sampling valve.

Summary

Quantitative analysis is one of the more important uses for gas chromatography. It relies primarily upon peak area measurement, although there are occasions when peak height measurement is a satisfactory (or even the preferred) alternative. Area measurement is now best performed by an integrator, although manual methods still have a place. Quantitation is rarely carried out by simple calibration with standards (except in the case of gas analysis), but is done by area normalisation or the method of internal standards. Depending upon the particular analysis, relative precisions are in the order of 1% to 2.5% under ideal conditions. Errors are higher if peak shape or resolution is poor or if the conditions depart from the ideal in some other way.

SAQ 7a

Draw lines joining each of the analytical problems given below to the most appropriate method of quantitation:

(*a*) determination of ethanol in benzene	(*i*) area normalisation
(*b*) determination of argon in a welding gas atmosphere	(*ii*) direct calibration to produce a simple calibration graph
(*c*) determination of all the components in a mixture of hydrocarbons used as a paint solvent	(*iii*) internal standard

SAQ 7b

Select from the following list the properties which you would look for in an internal standard for gas chromatography and place them in order of decreasing importance (ie the one to which you give highest priority at the top end and the one to which you give lowest priority at the bottom).

Low volatility, similar chemical composition to the analyte(s), chemical and thermal stability, similar relative molecular mass to the analyte(s), different retention time to the analyte(s), similar relative response factor to the analyte(s), elutes approximately mid-way between the analytes, different retention time to all components of the mixture.

Learning Objectives

After studying the material in Part 7, you should now be able to:

- discuss the sources of error in quantitative gc;
- describe the method of internal standardisation, as used in gc;
- describe methods suitable for the quantitative analysis of mixtures of gases by gc.

Self Assessment Questions and Responses

SAQ 1a

Indicate, by circling either T for True or F for False, whether you agree with each of the following statements:

(*i*) The best way to analyse a mixture of H_2O and D_2O is by glc.

T / F

(*ii*) If a mixture of ethane and ethene is not separating very well with argon as the carrier gas, it can be improved by switching to hydrogen.

T / F

(*iii*) If you want to analyse a mixture of 1,2-, 1,3- and 1,4-dimethylbenzenes you are likely to have to use a capillary column.

T / F

(*iv*) If you wanted to identify the contents of an unlabelled bottle containing a mixture of organic solvents that has been found on the street, you would use gc.

T / F

Response

(*i*) The correct response was F; there is a slight difference in volatility because of the isotopes, but it is not enough to make separation easy. It would be better to use a spectrometric technique.

(*ii*) Correct answer F; molecules are so far apart in the gas phase that they do not have significant intermolecular attractions. The nature of the carrier gas therefore has very little effect on the distribution equilibrium and hence on separation.

(*iii*) T is correct. although there is enough difference in the boiling points of 1,2-dimethylbenzene and the other two for separation, the difference between 1,3-dimethylbenzene and 1,4-dimethylbenzene is very small and ordinary packed columns do not separate them unless some rather exotic stationary phases are used. You need the efficiency of a capillary column to do this easily.

(*iv*) You could argue about this one, but on balance I would say F was correct. Gc might tell you how many compounds were present, but not what they were, unless the smell gave you a hint so that you knew what standards to compare them with. Molecular spectrometry would give you a much better idea of what sort of compounds were present. In the end, you would probably use a combination of both.

****

SAQ 1b

Indicate, in the space provided, the technique, chosen from the list A to E, that would be appropriate for:

(*i*) Distinguishing between samples of polyethylene and polypropylene

(*ii*) Analysing a mixture of phenols ⟶

SAQ 1b (cont.)

(*iii*) Estimating polychlorinated biphenyls (PCB's) in infertile sparrowhawks' eggs

(*iv*) Analysing a mixture of petrol and diesel fuel

A = pyrolysis gas chromatography;
B = use of a selective detector;
C = derivatisation;
D = temperature programming;
E = preparative gc.

Response

(*i*) Answer = A; these polymers are not sufficiently volatile for gc. They would have to be degraded to smaller molecules which would be volatile enough, by pyrolysis. Derivatisation, even if it were possible for these polymers, would not cause a big enough improvement in their volatility.

(*ii*) Answer = C; phenols do not chromatograph very well. They have high boiling points and the hydroxyl group causes them to 'tail', or give very badly shaped peaks. Converting them to methyl or silyl derivatives overcomes both problems.

(*iii*) Answer = B; there would be a lot of other organic matter present which it would be difficult to separate from the PCB's. Best to avoid the difficulty by using a detector which responds selectively to chlorinated compounds and so ignores all the other 'junk'.

(*iv*) Answer = D; with the very wide range of boiling points in this mixture, you would find it difficult to find a single temperature suitable for all the components present. Hence temperature programming should be used.

****

SAQ 2a

The sixth form at your local school has built a gas chromatograph, illustrated below. It is used to separate dichloromethane, trichloromethane and tetrachloromethane. As each of these components emerges, the flame turned from colourless to a typical *copper blue/green.*

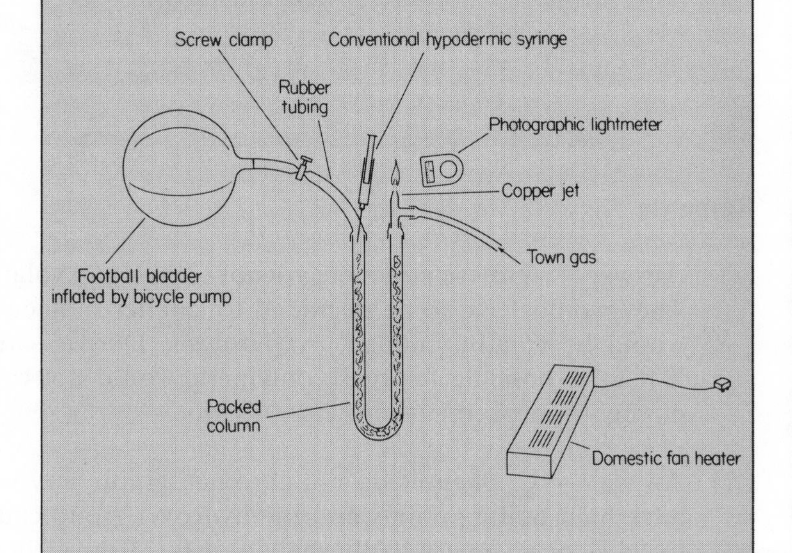

List at least six points at which this design departs from that of an *ideal* gas chromatograph.

Response

Your list could include any of the following:

(1) Oxidising carrier gas (air)

(2) Flow rate not constant (both due to random fluctuations and a decrease as the pressure in the bladder falls).

(3) Flow rate not measured.

(4) Syringe will be neither accurate nor reproducible for small volumes of liquid.

(5) No injection heater.

(6) Poor, or non-existent, temperature control.

(7) There is no wavelength filter so that any change in light intensity (due either to sudden sootiness in flame or change in background lighting) will interfere.

(8) Detector's response is probably far from linear.

(9) Detector will only detect chlorinated compounds.

(10) There is no permanent record of results.

Answers 1 to 7 are probably the most important defects in this instrument. If you included at least five of these in your list, well done. You seem to have got the idea and are thinking critically about gas chromatography. If you missed many of them, perhaps it might be an idea to have another look at Part 2.

****

SAQ 2b

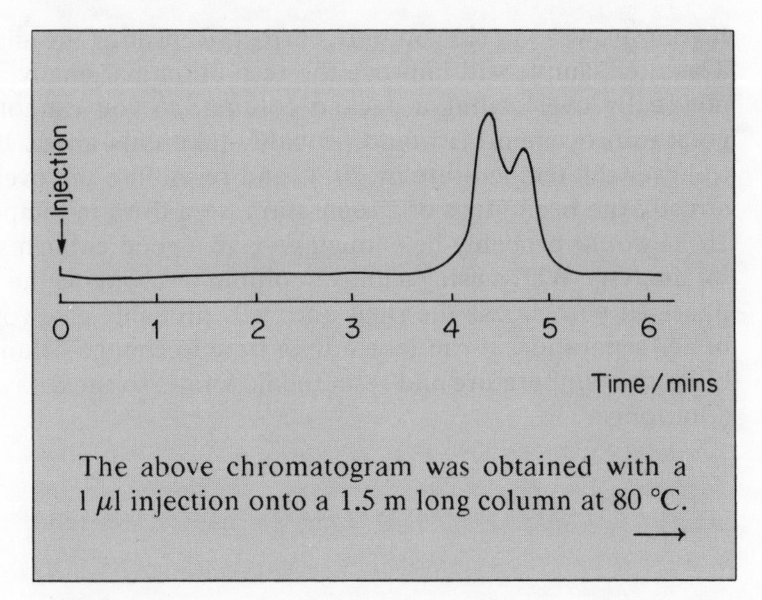

The above chromatogram was obtained with a 1 μl injection onto a 1.5 m long column at 80 °C.

⟶

SAQ 2b (cont.)

In order to perform a quantitative analysis for the two components, the resolution of the peaks must be improved. To do this, and perform the analysis *with the minimum delay*, would you:

1. repeat the analysis with a smaller sample at a higher sensitivity?
2. lower the temperature and repeat the analysis?
3. repeat the analysis using a longer column of the same stationary phase at the same temperature?
4. repeat the analysis using a different stationary phase?

Response

If your answer was (2) you were correct. Repeating the analysis with a smaller sample will improve the resolution marginally, but 1 μl is not really overloading a packed column, so you cannot expect a great improvement. It would probably take only about 10 minutes to lower the temperature by 20 °C and restabilise the oven; there is already the beginnings of a separation, so a drop in temperature of 20 °C would probably be enough go give a good enough separation for analysis. Whilst using a longer column or changing the stationary phase (if you choose the right one) will probably give you an even better separation, it can take a long time to change columns, restabilise the temperature and reset the flow rate, so the delay would be a lot longer.

****

SAQ 2.5a

Describe a device for accurately injecting 0.05 μl of a solution of benzene in propanone into a gas chromatograph, in order to carry out a quantitative analysis.

Response

Sorry about the trick question, but it's my way of emphasising the point that there *is* no device readily available which can do this! Congratulations if you avoided the trap. If you fell into it, you probably described a standard 'plunger in needle' microsyringe, which is what you would be most likely to use. It is not accurate, though, when handling such small volumes, particularly if the sample is volatile. For a bonus point, how would you overcome this problem?

If you suggested using an internal standard, you were correct. It would allow for inaccuracies in the syringe, and, if it were chosen to be as similar to benzene as possible it would also allow for losses by evaporation.

****

SAQ 2.6a

Indicate the characteristics (there may be more than one in each case), chosen from the list A to D, which you associate with the detectors in the list (*i*) to (*iii*):

A. Excellent linearity

B. Good selectivity

C. Sensitivity →

SAQ 2.6a (cont.)

D. Universality

(*i*) Electron capture detector

(*ii*) Flame ionisation detector

(*iii*) Katharometer

Response

(*i*) If you answered (B) then you have chosen the most obvious answer. It is its selectivity for such things as halogenated organics which is responsible for many of its applications. If you have chosen (C), you can also make a good case for your answer, as it is probably the most sensitive of the three detectors. But it is most certainly not linear (A) or universal (D). If you chose either of these, perhaps you should re-read Section 2.6.3.

(*ii*) I think I would have chosen (A) first, and then (C). Few detectors are as linear as the FID and it is very sensitive although there are several detectors which are at least equally sensitive.

I think the position over (B) and (D) is more debatable. Although it is almost universal for organics, failing to detect only a few organic gases, it does not detect inorganics. Whilst it is true that many laboratories are only concerned with the analysis of organics, and would regard the FID as almost universal, this is rather a narrow view. Again, it is selective for organics as opposed to inorganics it is certainly not selective *within* organics. You might argue the case for (B). It would be much more difficult to argue the case for (D) and you certainly could not have them both!

(*iii*) (D) is only satisfactory answer. It is not selective or sensitive and the linearity is only moderate.

****

SAQ 2.6b

Fill in the blanks in the following paragraph, choosing the missing words from the list below:

If you were asked to analyse the headspace gases in a storage tank containing gasoline, the detector you would choose would depend upon why it was being analysed. If it was only the relative proportions of the individual hydrocarbons in the vapour phase which was needed, you would use a/an............................., because it is suitably.......................for hydrocarbons, but if what was needed was to know if air had been excluded by purging with nitrogen, you would use a/an................... because it is................

Electron capture detector

Flame ionisation detector

Katharometer

Linear

Selective

Universal

Sensitive

Response

If you were asked to analyse the headspace gases in a storage tank containing gasoline, the detector you would chose would depend upon why it was being analysed. If it was only the relative proportions of the individual hydrocarbons in the vapour phase which was needed, you would use a *flame ionisation detector*, because it is suitably *sensitive* for hydrocarbons, but if what was needed was to know if air had been excluded by purging with nitrogen, you would use a *katharometer*, because it is *universal*.

****

SAQ 2.6c

Would the following analyses be successful or unsuccessful because the named detector is inappropriate?

(*i*) the determination of traces of methylbenzene in benzene, using a katharometer.

(*ii*) the determination of carbon monoxide and carbon dioxide in furnace flue gases, to determine the efficiency of combustion, using a katharometer.

(*iii*) the determination of carbon monoxide and carbon dioxide in furnace flue gases, to determine the efficiency of combustion, using a flame ionisation detector.

(*iv*) the determination of both hydrogen cyanide and toluene diisocyanate in the products of combustion of a polyurethane foam, using a katharometer.

(*v*) the determination of both hydrogen cyanide and toluene diisocyanate in the products of combustion of a polyurethane foam, using a flame ionisation detector. ⟶

SAQ 2.6c (cont.)

(*vi*) the determination of perchloroethylene (dry cleaning fluid) in a petroleum extract from clothing using an electron capture detector.

Response

(*i*) Would be *unsuccessful*, since the katharometer lacks the sensitivity needed for trace analysis. If 10 ppm of methylbenzene were present in the benzene, you would need to inject about 100 μl of the benzene to get a measurable methylbenzene peak.

(*ii*) Would be *successful*, the katharometer being able to detect both these gases.

(*iii*) Would be *unsuccessful* since the FID is unable to detect either of these gases.

(*iv*) Would be *successful*, although the katharometer would be struggling to detect the small quantities of hydrogen cyanide present. A quite large sample volume will have to be used. The katharometer is capable of detecting both compounds.

(*v*) Would be *unsuccessful*. The FID does not detect hydrogen cyanide and so the analysis would fail.

(*vi*) Would be *successful*. Because the ECD is selective for halogenated compounds, it will give a large peak for the perchloroethylene and will not detect the hydrocarbons of petroleum which would otherwise give a jumble of peaks which could well interfere.

How did you get on? If you got more than four of them right, you have a good enough grasp of the problems of choosing a detector to go ahead. If you got four or less right, perhaps you should have another quick re-read of Section 2.6 before going on.

****

SAQ 3.2a

Name two materials commonly used for the tubing from which packed gas chromatography columns are made, and for each of them name two good points and one bad point.

1. ____________ Good points ____________

____

Bad point ____________

2. ____________ Good points ____________

____

Bad point ____________

Response

There are only two materials commonly used; each of them has more than two good points and one bad point, so your answer should have selected from the more extensive list I have shown

1. Glass

Good points: Chemical inertness
Thermal stability
Very low adsorptive capacity
Transparency

Bad point: Fragility
Slight adsorptive capacity

2. Stainless Steel

Good points: Chemical inertness
Thermal stability
Strength

Bad point: Occasional catalytic decomposition

If you are too far out, say more than one or two wrong, perhaps it might be a good idea to re-read 'Materials' which appears at the end of Section 3.2 of this Unit.

****

SAQ 3.2b

Indicate, by circling either T for True or F for False, whether you agree with the following statements:

1. The most useful size of packed column in gas chromatography is about 1.5 m long and 4 mm id.

   T / F

2. The development of capillary columns has rendered packed columns obsolete.

   T / F

3. Since small particles of packing give so much better performance than large ones, it is better to use a packing which has a small average particle size, even if its particle size range is much larger than an alternative packing which has a somewhat large average particle size and a much smaller particle size range.

   T / F

Response

1. True, though there are supporters of the 2 mm id column, since it uses less packing material. There will, of course, be occasions when a longer column will be needed for a pair of components that are difficult to separate or when a shorter column will be needed to keep retention times within acceptable limits.

2. False. You only have to look in the catalogue of any chromatographic supplier. Capillary columns do give far better resolution, but they are expensive and fragile and for some of them the problems of injecting and detecting the necessarily small samples are not easy to overcome. Finally, of course, you would look a bit silly trying preparative gas chromatography on a capillary column.

3. False. The main benefit of using small particles is the improvement in evenness of packing, and this is lost if a much wider particle size range packing is used. The different sized and shaped particles pack very unevenly.

****

SAQ 3.3a

Place the following stationary phases in order of increasing polarity:

polyethylene glycol succinate (PEG-S),

polyethylene glycol – relative molecular mass 400 (PEG 400),

polyethylene glycol – relative molecular mass 20,000 (PEG 20M),

hexadecane,

tritolyl phosphate,

polypropylene glycol adipate (PPGA).

Response

Hexadecane < tritolyl phosphates < PEG 20M < PEG 400 < PPGA < PEG-S

PEG-S is the most polar, consisting mainly of ester carbonyl groups, PPGA is less polar, the carbonyl groups being 'diluted' by the extra methylene groups, PEG 400, the polyether, is the next most polar, being more polar than PEG 20M because it has a higher proportion of terminal hydroxyl groups because of its lower relative molecular mass. Tritolyl phosphate is less polar again since the three aryl groups 'dilute' the polar phosphate group and the alkane, hexadecane, is least polar.

****

SAQ 3.3b

Benzene, cyclohexane and ethanol are to be separated by glc. Given that they all boil between 70 °C and 80 °C, indicate by circling either T for True or F for false whether you agree with either of the following statements.

1. On a squalane stationary phase at 70 °C the order of elution would be:

   ethanol, followed by benzene, followed by cyclohexane.

   T / F

2. On a polyethylene glycol succinate (PEG-S) stationary phase at 70 °C, ethanol would elute after both benzene and cyclohexane.

   T / F

   You might like to consider what would happen if PEG 400 were used as a stationary phase instead.

Response

1. If your response was T, well done. You obviously realised that with such similar boiling points, the separation will be solely due to differences in solubility.

If your answer was F, sorry. We have something to sort out. With such similar boiling points, the separation will be solely due to differences in solubility. Squalane will not form hydrogen bonds or dipole attractions. The only forces left to encourage solubility are dispersion forces which are strongest between similar molecules, ie the two alkanes, squalane and cyclohexane, and weakest between dissimilar molecules, ie squalane and ethanol. The order of elution in the statement is therefore correct.

2. If you answer was T, we are going to have to work on it, because it was WRONG. This is because benzene and ethanol elute together, after cyclohexane. The weak hydrogen bonding between ethanol and the polyester PEG-S is swamped by the big dipole–dipole attractions between the carbonyl of the polyester and the hydroxyl of the alcohol. This is matched by the large dipole-induced dipole attraction between the carbonyl of the polyester and the dipole it induces in benzene by polarising its π-electrons. Hence their similar solubility.

If your answer was **F**, well done.

You might like to consider what would happen if PEG 400 were used as stationary phase instead.

Answer to supplementary question:

If you suggested that ethanol would elute AFTER benzene, you were right. You realised that ethanol would now form much stronger hydrogen bonds and that benzene would form much weaker dipole-induced dipole attractions as a result of the smaller dipole and greater hydrogen bonding ability of PEG 400. Ethanol would now be more soluble than benzene and so elute after it.

****

SAQ 3.3c

For the three mixtures below, select in each case the answer (*a*), (*b*) or (*c*) which you think is the correct order in which the named components of the mixture will elute on the given stationary phase.

1. Cyclohexane (bp = 81 °C) and cyclohexene (bp = 83 °C) on dinonyl phthalate (DNP).

   (*a*) More or less together
   (*b*) Cyclohexane then cyclohexene.
   (*c*) Cyclohexene then cyclohexane.

2. Methoxybenzene (anisole, bp = 154 °C) and 1-methylethylbenzene (cumene, bp = 152 °C) on polyethylene glycol (PEG 400).

   (*a*) More or less together.
   (*b*) Anisole then cumene.
   (*c*) Cumene then anisole.

3. Hexane (bp = 68 °C), 1-methyl-1-(1-methylethoxy) ethane (diisopropyl ether) (bp = 68 °C) and propan-2-ol (bp = 83 °C), on polyethylene glycol succinate (PEG-S)

   (*a*) Hexane, diisopropyl ether, then propan-2-ol.
   (*b*) Diisopropyl ether, hexane, then propan-2-ol.
   (*c*) Hexane, propan-2-ol, then diisopropyl ether.

Response

1. Correct answer = (*b*). DNP is slightly polar, in spite of its very

high proportion of alkyl and aryl groups. The carbonyl groups can induce a small dipole in cyclohexane by polarising the π-electrons. This will increase cyclohexene's solubility in the stationary phase, which, together with its slightly lower volatility, will lead to it eluting after the completely unpolarisable cyclohexane.

If you answered (*a*) I guess that you missed the polarisability of the π-electrons, but if you answered (*c*) I am at a loss to know what you were thinking. If you have read the discussion above and cannot see where you went wrong, then you probably need to discuss this with your tutor.

2. Correct answer = (*c*). PEG 400 has a high proportion of hydroxyl groups which can act as proton donors in hydrogen bonding with the oxygen atom of the ether. PEG 400 does not have a strong enough dipole to induce a dipole in the aryl rings of either anisole or cumene by polarising the π-electrons. Consequently, the ether is more soluble than the hydrocarbon. Combined with its slightly higher boiling point this means that the ether migrates more slowly. Incidentally, had we chosen PEG-S as the stationary phase, because it is a much poorer proton donor in hydrogen bonding and also has a much greater dipole, we would not have separated the mixture nearly as well.

If you answered (*a*) you probably missed the difference in hydrogen bonding ability. Again, there is no obvious reason for answering (*b*), so if you did this you ought to find someone to discuss it with, unless the above comments have made you see where you were going wrong.

3. Correct answer = (*a*). The propan-2-ol is less volatile and has a greater solubility in the stationary phase due to its dipole and hydrogen bonding ability (as a proton donor). The ether has a lesser dipole and will not hydrogen bond with PEG-S as neither of them has a hydrogen atom. Hexane has no dipole and no hydrogen bonding ability.

****

SAQ 3.3d

Explain, in the space below, why gases are more likely to be analysed by gsc than glc (and if you just say 'because gsc separates gases better than glc' I shall ask you why it does, so you might as well get down to fundamental reasons to begin with)!

Response

Your explanation should have included mention of the fact that gases are not sufficiently soluble in most liquids at 50 °C and above for them to be retained significantly by glc so that very long columns and sub-ambient temperatures would have to be used whereas gases are strongly adsorbed on many solids. Consequently retention would be reasonable on short columns above room temperature. Separation will be possible. You should also have mentioned the greater selectivity of adsorption due to the rigid geometry of adsorbents as well as molecular sieving, both of which will be necessary to separate gases which may have very little difference in polarity.

Your answer may have been less detailed (I am interested in correcting misunderstandings, so have to say more than you), but if you did not get substantially the above answer, perhaps you should re-read 'Solid Stationary Phases' in Part 3 of this Unit.

****

SAQ 3.3e

If you were asked to determine the concentration of carbon monoxide and carbon dioxide in a boiler flue, which of the following stationary phases would you use?

1. To determine carbon monoxide __________

2. To determine carbon dioxide ___________

Stationary phases:

Alumina, silica gel, graphitised carbon black, Linde molecular sieve, porous polystyrene (Porapak).

Response

You may have noticed that all the stationary phases suggested to you are adsorbents. Liquid stationary phases are generally of little use for separating non-hydrocarbon gases unless sub-ambient temperatures and very long columns are used.

1. A Linde molecular sieve is probably the best phase for separating carbon monoxide from the other gases in air (see Section 3.3 of this Unit). Porous polymers are not generally very good for separating permanent gases, being, like graphitised carbon black, better for organic liquids. Alumina and silica gel do not separate carbon monoxide from oxygen or nitrogen.

2. Silica gel would be needed to separate carbon dioxide. You could not use a molecular sieve because carbon dioxide is irreversibly adsorbed, which makes the analysis slightly tedious – you have to inject onto two separate columns.

****

SAQ 3.4a

List three respects in which diatomaceous earth falls short of the ideal as a support for gas chromatography.

1.

2.

3.

Response

Diatomaceous earth is probably the best support we have, but it is far short of ideal. In particular:

1. It retains some adsorptive capacity.

2. It is physically fragile and so breaks up into smaller particles leading to a wider particle size range.

3. Its particles are irregular in shape.

4. It has deep pores which fill with liquid stationary phase. Mass transfer can be slow as molecules diffuse down into them and then back up to the surface of the stationary phase.

I expect that you got the first three. The last one was not spelt out in the text, but was left to see if you were thinking about the implications of what you were reading, instead of just reading automatically.

****

SAQ 3.4b

On which of the following columns is the *tailing* of a phenol peak likely to be worst?

1. 10% Apiezon L on diatomaceous earth.

2. 2% Apiezon L on diatomaceous earth.

3. 10% PEG 400 on diatomaceous earth.

4. 3% PEG 400 on PTFE beads.

Response

The answer is No. 2 (2% Apiezon L on diatomaceous earth). Tailing of polar components is due, you will remember, to adsorption onto supports which are not entirely inactive. Adsorption isotherms are not often linear and the result is tailing. Well, PTFE is inactive, and PEG 400 is very polar so that it will block the active sites of adsorption on the diatomaceous earth, minimising adsorption of the phenol. Apiezon L is a hydrocarbon grease and will be only weakly adsorbed so that it will not block the active sites in competition with phenol. Thus the choice for *the column which gives the worst tailing* lies between 1 and 2. The difference between them is that with so much more liquid stationary phase in 1 (10% instead of 2%) simple partition, which does not give rise to tailing since partition isotherms are usually linear over a wider range will play a much larger part whereas adsorption will be more important in 2. This was discussed at the end of Section 3.4 of this Unit.

****

SAQ 4a

A retention time of 2.33 minutes was obtained for ethanol, when it was chromatographed under the following conditions:

Sample size:	0.1 μl
Column:	1.5 m × 4 mm id; 10% PEG 400 on 100-120 mesh diatomaceous earth
Temperature:	70 °C
Flow rate:	40 cm^3 min^{-1}
Injection heater:	90 °C

Select, from the table of retention times (*a*) to (*g*), the value you would expect to get when the above conditions were modified by the changes detailed in (*i*) to (*vi*). In each case all other conditions remain unchanged.

(*i*) Flow rate = 55 cm^3 min^{-1}.

(*ii*) Injection heater: off.

(*iii*) Column 1.5 m × 4 mm id; 5% PEG 400 on 100–120 mesh diatomaceous earth.

(*iv*) Column: 2.7 m × 4 mm id; 10% PEG 400 on 100–120 mesh diatomaceous earth.

(*v*) Column: 1.5 m × 4 mm id; 10% Squalane on 100–120 mesh diatomaceous earth.

(*vi*) Sample size = 1.0 μl. ⟶

SAQ 4a (cont.)

(*a*) 0.24 min (*b*) 1.28 min (*c*) 1.71 min

(*d*) 2.35 min (*e*) 3.10 min (*f*) 4.23 min

(*g*) 6.05 min

Response

(*i*) Correct answer: 1.71 min. Increasing the flow rate reduces the retention time, so I hope that you didn't choose (*d*), (*e*), (*f*) or (*g*). To a very reasonable approximation, the retention time is inversely proportional to the flow rate.

Therefore, at 55 $cm^3\ min^{-1}$,

$$\text{retention time} = 2.35 \times 40/55 \text{ min}$$

$$= 1.71 \text{ min}$$

(*ii*) Correct answer: 2.35 min. Changing the injection heater temperature does not alter the distribution equilibrium at all. It only alters the rate of vaporisation of the sample on injection. Therefore, turning the heater off will lead to broader peaks but will not alter the retention time.

(*iii*) Correct answer: 1.28 min. If you chose this you realised that reducing the amount of stationary phase would move the distribution equilibrium in favour of the mobile phase and reduce the retention time. If you halve the amount of stationary phase, you will halve the time the analyte spends in it. To a first approximation this will halve the retention time. (If you are interested, the 'discrepancy' arises because you will not reduce the time that the analyte spends in the mobile phase, which would be less than a quarter of a minute in this column. Your new retention time would therefore be approximately 0.25 min plus half of 2.15 min).

(*iv*) Correct answer: 4.23 min. I expect that you saw that the column length was increased, and that because that would lead to increased retention times you need only consider answers (*e*), (*f*) and (*g*). If you increase the column length, you will increase the time spent in both mobile and stationary phases in proportion to the increase. The new retention time would be equal to the old retention time multiplied by 2.7/1.5. Hence the answer of 4.23.

(*v*) Correct answer: 0.24 min. I hope that you didn't think this too unfair; congratulations if you got it right.

To get the right answer you will have had to remember that squalane is non-polar and will hardly dissolve ethanol at all at this temperature (which is revision of the previous Part!). The retention time of ethanol will therefore be a lot less than 2.35 min. In fact it will be hardly more than the time that the carrier gas would take to pass through the column. The dimensions of the column mean that it has a volume of about 10 cm^3, so at 40 cm^3 min^{-1} the carrier gas would take less than a quarter of a minute to pass through the column (you would have to allow for the volume taken up by the packing). So, of the answers available, (*a*) looks most appropriate.

(*vi*) Correct answer: 2.33 min. Of course you remembered that retention times are to all intents and purposes independent of sample size!

****

SAQ 4b

The series of chromatograms (*a*) to (*d*) were obtained with successive injections onto a 1.5 m column packed with 10% dinonyl phthalate, at a temperature of 100 °C and a flow rate of 40 cm^3 min^{-1}. The samples were taken from the atmosphere of a badly ventilated paint spraying booth, about which complaints had been made. The analyst's technique is faulty. Can you spot his mistake. ⟶

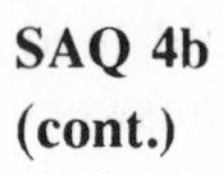

SAQ 4b (cont.)

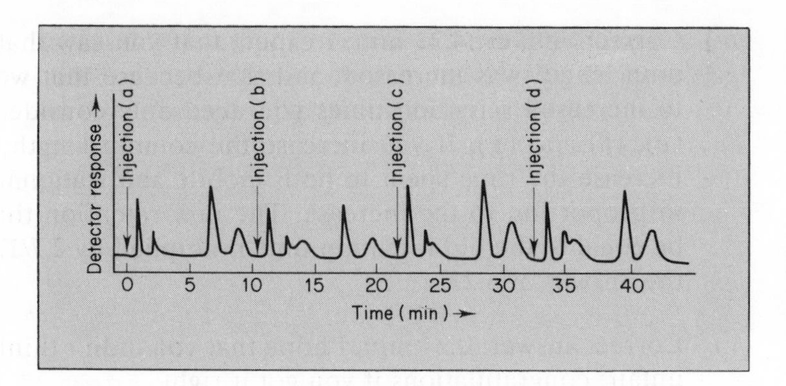

Response

If you spotted the overlapping injections caused by his impatience – well done. The clue was the rather broad third peak in (*b*) to (*d*), which seems to have a variable retention time. In fact, it is really the fifth peak of (*a*) to (*c*) turning up in the later chromatograms because they were injected too soon.

****

SAQ 4c

A 0.1 μl sample of ethanol injected into a gas chromatograph fitted with a flame ionisation detector, gave a peak 9 mm high at an attenuation of $\times 10^5$

(*i*) How high a peak would be given by 1 μl of an aqueous solution containing 100 mg dm^{-3} at an attenuation of $\times 10^2$?

(*ii*) What attenuation would you use if you wanted to inject 0.1 μl of a 1% aqueous solution of ethanol, if the chart paper is 25 cm wide?

Response

The first thing to remember, and I am sure that you did, is that the FID has very good linearity.

(*i*) Answer: approximately 9 cm. well done, if you got it. If not, perhaps you should read the following explanation.

1 μl of a 100 mg dm^{-3} soln. contains

$$100 \times \frac{1}{1\,000\,000} \text{ mg EtOH}$$

ie if you assume a density of 1 g cm^{-3} for ethanol, approximately 10^{-4} μl EtOH

10^{-1} μl at $\times\ 10^5$ gave a 9 cm high peak

10^{-4} μl at $\times\ 10^5$ would give a $\dfrac{9 \times 10^{-4}}{10^{-1} \text{ cm}}$ high peak

ie 9×10^{-3} cm high

10^{-4} μl at $\times\ 10^2$ would give a $\dfrac{9 \times 10^{-3} \times 10^5}{10^2 \text{ cm}}$ high peak

ie 9 cm high

(*ii*) Answer: between 5×10^2 and 5×10^3. Again, well done if you got it right, but read on to find out where you went wrong, if you did.

0.1 μl of ethanol at $\times 10^5$ gave a peak 9 cm high

0.1 μl of a 1% solution at $\times 10^5$ would give a peak 9/100 cm high

ie 0.09 cm high.

I would not like to use a peak less than 1 cm high, so the attenuation would need to be reduced by a factor of about 20 (which would give a peak 0.09 × 20 cm, ie 1.8 cm high). The attenuation would therefore need to be reduced to at least $10^5/20$, ie 5×10^3.

On the other hand, peaks larger than 25 cm will go off scale. If we reduced the attenuation by more than 200, our 0.09 cm peak would become more than 0.09 × 200 cm, ie 18 cm high and it would be getting close to being too large. The attenuation could not be reduced to less than $\times 10^5/200$, ie 5×10^2.

I hope that you have got the idea by now. You need to have a rough idea of the size of the peaks to expect, or else you can waste a lot of time waiting at too high an attenuation for peaks to appear, when in fact they have already emerged – but too small for you to see! If you are still unhappy, re-read Section 4.6.

****

SAQ 4d

The three chromatograms below are the result of three successive, identical injections of the same sample. What has gone wrong, and what should be done to correct it?

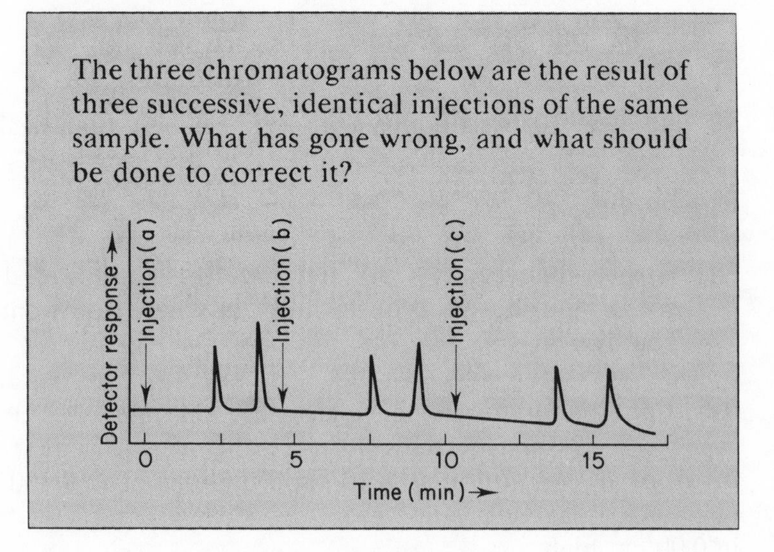

Response

Did you spot the 'leaky septum syndrome'? Congratulations if you did. If you missed it, the clues were the falling baseline, the increasing retention times and reduction in peak heights. There are really only two things which could cause this – either a progressive fall in temperature or a progressive fall in flow rate. Ovens are pretty reliable, so a fall in flow rate is more likely, and a leaky septum is the most likely cause of this. Check it and replace it.

****

SAQ 4e

The following chromatograms were obtained under the conditions appended to each one. In each case, what changes would you make in order to improve them and make them usable for the analysis of the relevant mixture?

(a)
Sample: 0.1 μl of benzyl bromide (b.p.t. = 198 °C) + 1,2-dibromobenzene (b.p.t. = 244 °C)
Column: 1.5 m x 4 mm; 10% squalane on diatomaceous earth
Temperature: 140 °C
Flow rate = 40 cm^3 min^{-1}
Injection Heater = 175 °C
Attenuation = x 10^5
Detector = FID

Detector response → Injection
0 5 10 15 20 25
Time (min) →

(b)
Sample: 10 μl of ethanol + propan-1-ol
Column: 0.75 m x 4 mm; 10% PEG-400 on diatomaceous earth
Temperature: 100 °C
Flow Rate: 40 cm^3 min^{-1}
Injection Heater: 105 °C
Attenuation: x500
Detector: TCD (Katharometer)

Detector response → Injection
0 1 2 3 Time (min) →

(c)
Sample: 5 μl of benzene + methylbenzene
Column: 1.5 m x 4 mm; 10% DNP on diatomaceous earth
Temperature: 75 °C
Flow Rate: 35 cm^3 min^{-1}
Injection Heater: off
Attenuation: x 200
Detector: TCD (Katharometer)

Detector response → Injection
0 1 2 3 4 Time (min) →

⟶

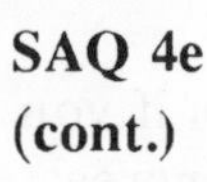

SAQ 4e (cont.)

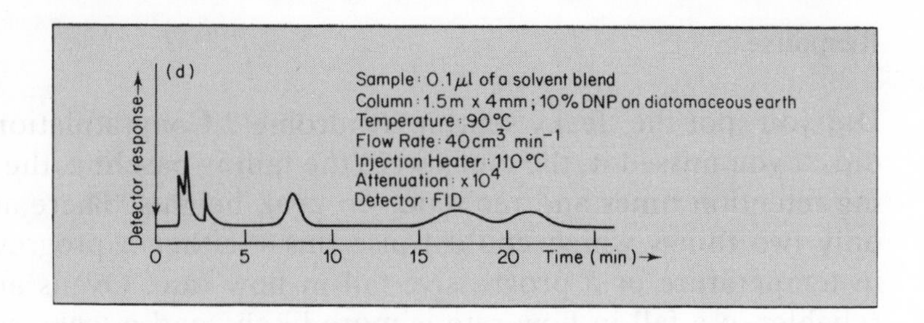

Response

(*a*) If you said 'Change the column' you wouldn't be wrong – that can cure most ills. Only, I would immediately ask 'What to?', which puts us almost back where we started.

Let us define the problem. The resolution is adequate, but the retention times are too long and the peaks too small. We need to reduce the retention time, perhaps at the cost of the resolution, and increase the peak heights. This could be done in about four ways.

If you suggested *increase the flow rate*, I would agree that it was the simplest and quickest answer. 100 cm^3 min^{-1} is quite feasible, would more than halve the retention times, forcing the peaks to be narrower and higher and would cause only a modest loss of retention.

If you suggested use a shorter column, again I would agree that it would work, but it might take a day to pack and condition a new, 50 cm long column. The alternative of using a 1.5 m column with a 2% stationary phase loading would also work, but would take a day to pack and condition.

If you suggested raising the temperature, I would NOT approve! You should have checked up on the stationary phase. 140 °C is the temperature limit of squalane, so if you raised it by the 40 °C necessary, you would see your stationary phase disappear up the chimney! It would cost you a new column. Of course you could raise the temperature if you changed to a stationary phase with a higher limit, eg Apiezon L which has similar characteristics to squalane, but a 350 °C limit. You are back to changing the column, which takes time.

Just reducing the attenuation would certainly increase the peak heights, but it would do nothing about the excessively long retention times. It is not really a satisfactory solution.

(*b*) The problem in this case is that the retention times are short and the resolution poor. We need to improve the resolution, and can afford to increase the retention times to do it.

If you suggested lowering the temperature, that would be a good start. At 100 °C it is well above the boiling points of both alcohols. Lowering it to 60 °C or 70 °C would probably do the trick without too much time lost.

If you suggested lowering the sample size and attenuation, I would agree that 10 μl is too large, but reducing it would not make a large enough difference on its own. It would, however, put the finishing touch to the improvement made by lowering the temperature, so I would suggest combining these two. The advantage is that there is little time penalty.

If you suggested increasing the column length, you have a feasible answer (75 cm is shortish). It would take a lot longer to put it into operation than the previous solution, so I would not choose it myself.

(*c*) The retention times are reasonable, the difference in retention times is enough, but the peaks are much too broad, given their retention times, for all to be well. The answer lies in looking for the reason for this.

Sample size is a little large, but not unreasonable, column length, temperature and flow rate are not unreasonable, which only leaves the injection heater. If you spotted that it was OFF, well done. You just need to turn it on and wait for it to warm up. (It might not have been so easy to spot that the heater element was burned out. This would have had the same effect).

The usual solution to all ills – changing the column – is not really satisfactory here. Benzene and methylbenzene are so similar in characteristics that changing to the less polar column which would normally be recommended for them will lead only to somewhat longer retention times and a slightly better resolution. The improvement would not be enough to overcome the disadvantage of those broad peaks. No – if you get such bad peaks shapes, something is wrong with the chromatography. The only answer is to find out what it is and to put it right.

(*d*) Well, I am sure that you spotted the poor resolution of the early components and broad peaks and long retention times of the later components which indicates a wide boiling range mixture. There is only one answer – temperature programming. You would need to choose the programme carefully, though, because of the poor resolution of the early peaks. If you started at 90 °C and raised the temperature at 12 °C min^{-1} you would make the resolution of these early peaks poorer and soon reach the temperature limit of DNP (150 °C). If you started lower (say 70 °C) and then raised it at 8 °C min^{-1} to 150 °C, the ten minutes it would take would probably be enough to elute the last components. If you suggested something like this, well done. We seem to be ready to go on to the next stage of the course. If you are not happy, it might be a good idea to re-read Part 4.

****

SAQ 5a

Would you expect the following analyses to require special treatment because of the low volatility of the sample?

(*i*) The fatty acids in a sample of soap.

Y / N

(*ii*) A sample of diesel oil.

Y / N

⟶

SAQ 5a (cont.)

(*iii*) The phenols used as raw materials for preparing phenolic resins

Y / N

(*iv*) A phenolic resin.

Y / N

(*v*) A light machine oil.

Y / N

Response

(*i*) Correct answer YES. Soap is a mixture of (mainly) the sodium salts of stearic acid ($C_{17}H_{35}COOH$) and palmitic acid ($C_{15}H_{31}COOH$). Their boiling points are too high for simple glc to be convenient.

(*ii*) Correct answer NO. Diesel oil, in spite of its name, is relatively volatile (boiling range = 250 °C to 400 °C) and it can be chromatographed quite well on a silicone column.

(*iii*) Correct answer NO. The phenol and cresols and other substituted phenols used boil between 180 °C and 240 °C, so they can be gas chromatographed quite easily. They are, however, quite likely to be derivatised in order to reduce the tailing to which they are very prone.

(*iv*) Correct answer YES. Obviously a solid resin will be too involatile for glc. You would use pyrolysis gas chromatography. On pyrolysis, phenolic resins yield mainly the phenol from which they were made together with the same phenol but with methyl groups attached to its phenyl ring where the methylene bridges were attached in the original resin. They would tail a bit, but it is still possible to get quite usable chromatograms.

(*v*) Correct answer YES. A light machine oil is not very volatile. You would need to use a high temperature stationary phase, a low stationary phase loading and a shortish column to get reasonable retention times.

****

SAQ 5b

Indicate, by circling T for True and F for False, whether you agree with the suggestion that the temperature limit of a column packed with 10% polyethyleneglycol adipate (PEG-A) on diatomaceous earth could be raised by:

1. Adding 0.5% phosphoric acid to the stationary phase.

   T / F

2. Coating the stationary phase onto PTFE beads instead of diatomaceous earth.

   T / F

3. Using carefully washed and silanized diatomaceous earth.

   T / F

4. Passing carrier gas through the column overnight at the temperature limit.

   T / F

Response

1. Correct answer F. As an acid, phosphoric acid would probably catalyse the decomposition of the stationary phase and lower the temperature limit. It is sometimes added to a polyester stationary phase when free fatty acids are being analysed, but this is done to reduce tailing, not to improve temperature limits.

2. Correct answer F. True, if the diatomaceous earth were catalysing the decomposition of the stationary phase, replacing it by inert PTFE would reduce decomposition. But if you look again at Section 3.4, you will see that we said that organic polymer supports tended to soften and clog the column, even at quite modest temperatures, so that the proposed course of action would be more likely to reduce the temperature limit.

3. Correct answer T. This is standard practice for any high temperature work.

4. Correct answer F. 'Conditioning' requires the column to be above the temperature limit – normally 25 °C above, but if you want to work at elevated temperatures you might want to condition at an even higher temperature.

****

SAQ 5c

Complete the following equation:

$$C_6H_5OH + CH_3{-}\underset{\displaystyle}{\overset{\displaystyle OSiMe_3\atop|}{C}}{=}N{-}SiMe_3 \rightarrow$$

How would you carry out this reaction?

Response

Your equation should have looked like this:

$$2\,C_6H_5OH + CH_3{-}\overset{\displaystyle OSiMe_3\atop|}{C}{=}N{-}SiMe_3 \rightarrow 2\,C_6H_5OSiMe_3 + CH_3CONH_2$$

You should have suggested adding bistrimethylsilylacetamide (BSA) to the acid in a screw capped vial, at room temperature and either leaving it at room temperature for a short time, or warming it at about 60 °C for a few minutes.

If you were uncertain about answering this question, perhaps you should re-read Section 5.3 of this Unit.

****

SAQ 5d

Complete the following paragraph by inserting the most appropriate word or phrase, chosen from the list given below, into the blank spaces.

The .... pyrolyser offers the most accurate control of the temperature of pyrolysis, but is probably more prone to .... than most, since it does not control .... On the other hand, the .... pyrolyser offers quite good control of the temperature of pyrolysis, but since it is also possible to achieve .... heating and cooling and .... timing of the duration of pyrolysis, it is probably less prone to secondary reactions.

Cooling	modest
Curie-point	overheating
duration of pyrolysis	secondary reactions
flash heater	slow
furnace	temperature.

Response

The FURNACE pyrolyser offers the most accurate control of the temperature of pyrolysis, but is probably more prone to SECONDARY REACTIONS than most, since it does not control DURATION OF PYROLYSIS. On the other hand, the CURIE-POINT pyrolyser offers quite good control of the temperature of pyrolysis, but since it is also possible to achieve RAPID heating and cooling and PRECISE timing of the duration of pyrolysis, it is probably less prone to secondary reactions.

****

SAQ 5e

Select the technique from (*i*) to (*iii*) which you think would be most appropriate for the gas chromatography of the following compounds:

1. D-Mannitol (a polyol found in mushrooms which acts as an antifreeze and prevents frost damage).

2. Perspex.

3. Dinonyl phthalate plasticiser.

4. Polystyrene.

5. Stearic acid ($C_{17}H_{35}COOH$)

(*i*) The use of a high temperature stationary phase.

(*ii*) Derivatisation.

(*iii*) Pyrolysis gas chromatography.

Response

1. Correct answer DERIVATISATION. The very high boiling point of mannitol is due to hydrogen bonding by its hydroxyl groups. Derivatisation overcomes this and, in fact, mannitol can be determined quite well by gas chromatography as its trimethylsilyl derivative.

2. Correct answer PYROLYSIS. Perspex is a polymer (polymethylmethacrylate), and as such will not be amenable to either high temperature stationary phases or derivatisation. It does depolymerise nicely on heating to even quite modest temperatures (200 °C–300 °C), so pgc works quite well.

3. Correct answer USE OF HIGH TEMPERATURE STATIONARY PHASES. Dinonyl phthalate is not very volatile (hence its use as both a plasticiser and a stationary phase), but it can be vaporised. There is no hydrogen bonding, so derivatisation will not improve the volatility, and attempts to pyrolyse it would probably lead to vaporisation without degradation. Pushing the temperature up by using a high temperature stationary phase thus offers the best hope of success.

4. Correct answer PYROLYSIS. Like perspex, polystyrene is a polymer and depolymerises nicely on heating. Pgc works well.

5. Correct answer DERIVATISATION. The hydrogen bonding of acids to form dimers in solution is well established. Forming a methyl or trimethylsilyl ester would overcome this and increase the volatility of the analyte, so making glc easier.

****

SAQ 6.2a

Which of the compounds in List A would you choose to use as an added standard compound in order to measure the relative retentions of the components of the mixture of chlorinated aromatic hydrocarbons whose chromatogram is shown below?

List A

Compound	t_R (min)
benzene	0.85
bromobenzene	3.9
2,4-dichloromethylbenzene	5.9
butyl benzoate	6.2

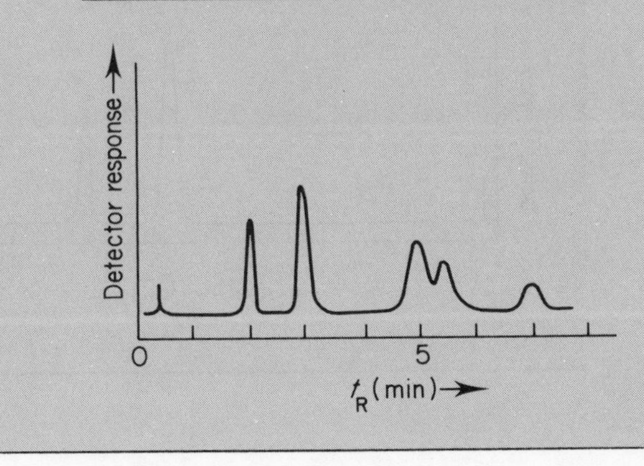

Response

The best choice is *bromobenzene*. From its retention time it looks as though it will be resolved from all components in the mixture and will elute roughly in the middle of the mixture. The relative retentions will be in the range 0.5 to 2 so that calculation and rounding up errors will be minimised. Benzene would elute far too early and any error in measuring its retention time would be magnified when it was divided into the other retention times to calculate relative retentions. Dichlorobenzene has a retention time which indicates that it would be likely to overlap with one of the components and butyl benzoate has far too high a retention time. On top of that, it is much more polar than the components in the mixture. Any slight change in temperature or the purity of the stationary phase could have different effects upon its retention and that of the mixture.

****

SAQ 6.2b

Calculate the Kovats Index of the compound eluting between decane and undecane on the chromatogram below.

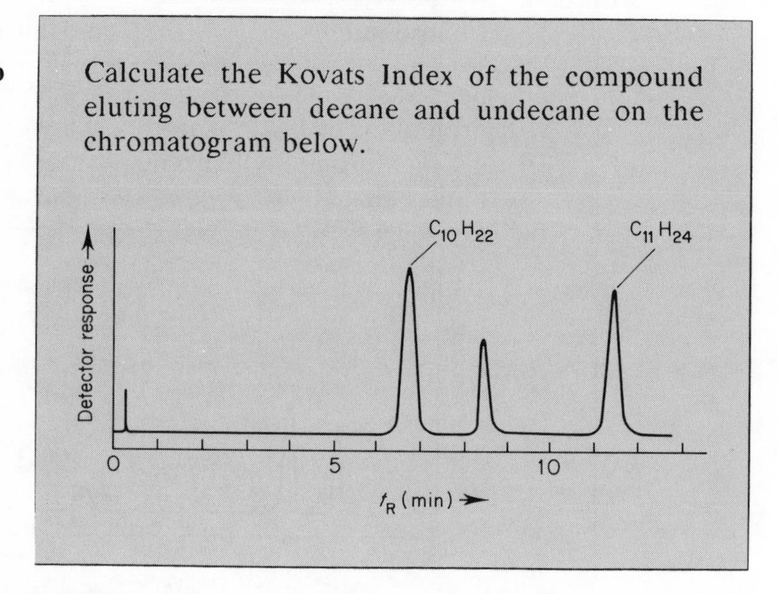

Response

The correct answer is 1043.

I calculated it as follows. First, I measured and tabulated the retention times and adjusted retention times of the significant peaks (using methane, which would not be significantly retained at this temperature, as the *air peak*).

Peak	t_R (min)	t'_R (min)
air peak	0.30	–
decane	6.72	6.42
unknown component	8.45	8.15
undecane	11.45	11.15

These data were then substituted into Eq. 6.1:

$$I = 100 \times \frac{\log 8.15 - \log 6.42}{\log 11.15 - \log 6.42} + 10$$

$$= 1043$$

∗∗∗∗∗∗∗∗∗∗∗∗∗∗∗∗∗∗∗∗∗∗∗∗∗∗∗∗∗∗∗∗∗∗∗

SAQ 6.3a

List three differences between analytical and preparative columns for gc.

1.

2.

3.

Response

Your answer should have included *wider id*, *large particle size packing* and *larger stationary phase loading*. All three are necessary to increase the capacity and maintain an adequate flow rate of carrier gas.

∗∗∗∗∗∗∗∗∗∗∗∗∗∗∗∗∗∗∗∗∗∗∗∗∗∗∗∗∗∗∗∗∗∗∗

SAQ 6.3b

(*i*) Which of the following traps would it be best to use for recovering diethyl phthalate (bp = 298 °C) in preparative gc?

1. a U-tube cooled in ice

2. a U-tube cooled in propanone/solid CO_2

(*ii*) Suggest a trap which would be more efficient than either.

Response

(*i*) The correct answer is 1. a U-tube cooled in ice. If you use the much colder refrigerant propanone/solid CO_2, the effluent gas stream is cooled too quickly,an aerosol is likely to result, and the droplets then pass right through the trap. Anyway, the vapour pressure of such a high boiling compound is very low at ice temperatures. Cooling to propanone/solid CO_2 temperatures is hardly necessary.

(*ii*) Packing the U-tube with any solid would increase the chances of aerosol droplets striking a solid surface and being trapped. Glass wool can be used for this purpose, but a column packing (20% silicone oil on diatomaceous earth) is much more effective.

If you had forgotten this, re-read Section 6.3.1 of this Unit.

****

SAQ 6.5a

Which of the following techniques can be said to be amenable to being usefully and effectively 'directly-linked' with gas chromatography so that spectra can be recorded without interrupting the chromatographic process (recorded 'on the fly')?

(*i*) infrared spectrometry

(*ii*) ultraviolet spectrometry

(*iii*) nuclear magnetic resonance spectrometry

(*iv*) mass spectrometry.

Response

The correct answers are (*i*) infrared spectrometry and (*iv*) mass spectrometry. They both give spectra which can be interpreted without too much ambiguity because of the large number of peaks. Ultraviolet spectrometry usually gives few peaks, and whilst it can give some indication of the type of compound present it is by no means as informative as ir or ms. It is doubtful if it is worth the effort necessary to link it directly. Nuclear magnetic resonance spectrometry would be very useful, but the practical difficulties are insurmountable at the moment.

****

SAQ 6.5b

Select from the following list those reasons for linking a gas chromatograph and a mass spectrometer which you think are valid and place them in a rank order (the most valid first).

⟶

SAQ 6.5b (cont.)

(*i*) to reduce costs

(*ii*) to check whether a peak on a gas chromatogram represents one component or two unseparated components

(*iii*) to improve quantitative analysis

(*iv*) to identify unknown peaks

(*v*) to improve detector sensitivity.

Response

My answer would be (*iv*) > (*ii*). The identification of unknown peaks is certainly the main reason for using gc/ms, but it is very useful to be able to scan the spectra of the *front* and *tail* of a peak to see whether they are identical (and therefore that the peak is homogenous, and presumably due to a single component) or different (and therefore the peak is due to two partly separated components). You could argue about reducing costs. Gc.ms is *very* expensive, by the time you have bought a separator and a computer, but there is really no other way to perform many of the analyses of which gc/ms is capable, so what can you compare its cost with? If there is no other way of doing it, then the alternative of having no result can be even more expensive.

It does not improve quantitative analysis in the generally accepted sense. Variable losses of the component in the separator mean that the mass spectral intensities will not be very reliable. Nor does it improve sensitivity greatly. The mass spectrometer is round about as sensitive as the FID.

****

SAQ 7a

Draw lines joining each of the analytical problems given below to the most appropriate method of quantitation:

(*a*) determination of ethanol in benzene	(*i*) area normalisation
(*b*) determination of argon in a welding gas atmosphere	(*ii*) direct calibration to produce a simple calibration graph
(*c*) determination of all the components in a mixture of hydrocarbons used as a paint solvent	(*iii*) internal standard

Response

Your answer should have looked something like this:

(*a*) determination of ethanol in benzene	(*i*) area normalisation
(*b*) determination of argon in a welding gas atmosphere	(*ii*) direct calibration to produce a simple calibration graph
(*c*) determination of all the components in a mixture of hydrocarbons used as a paint solvent	(*iii*) internal standard

The answer to (*a*) has to be the use of an internal standard. Area normalisation is not really suitable for trace analysis, since the peaks are of such different sizes, even if you do allow for different sensitivities by using relative response factors.

The answer to (*b*) would not be the use of an internal standard, because of the difficulty of adding it. It is unlikely to be area normalisation, although relative response factors can overcome the difficulty of the different sensitivity of the TCD for different gases and so make this method possible. Direct calibration using a gas sampling valve is the best choice.

(*c*) could be done by the use of an internal standard, but area normalisation would involve less work and would probably give equally good results. It might not even be necessary to use relative response ratios, since any detector will have quite similar sensitivities for a group of similar compounds such as these

****

SAQ 7b

Select from the following list the properties which you would look for in an internal standard for gas chromatography and place them in order of decreasing importance (ie the one to which you give highest priority at the top end and the one to which you give lowest priority at the bottom).

Low volatility, similar chemical composition to the analyte(s), chemical and thermal stability, similar relative molecular mass to the analyte(s), different retention time to the analyte(s), similar relative response factor to the analyte(s), elutes approximately mid-way between the analytes, different retention time to all components of the mixture.

Response

My list would look like this:

Chemical and thermal stability.

Different retention time to all components of the mixture.

Similar chemical composition to the analyte(s).

Elutes approximately mid-way between the analytes.

A standard which decomposes during the reaction is not a lot of use, so I put that first. After that, it is useless if it overlaps with other peaks, so that comes second. If it has a similar composition to the analyte, changes which affect losses of the analyte are likely to affect it in the same way, so that is desirable. Finally, if changes affect the analytes differently, an internal standard that elutes near to the middle of the mixture is likely to be affected in an intermediate manner so that one analyte does not suffer greater errors than the other.

The relative molecular mass and the relative response factors are irrelevent to the performance of the internal standard. Whilst a standard of low volatility might give more stable solutions and might be easier to weigh out, it is not going to be much use for gas chromatography as it will almost certainly give an unacceptably long retention time.

Units of Measurement

For historic reasons a number of different units of measurement have evolved to express quantity of the same thing. In the 1960s, many international scientific bodies recommended the standardisation of names and symbols and the adoption universally of a coherent set of units—the SI units (Système Internationale d'Unités)—based on the definition of five basic units: metre (m); kilogram (kg); second (s); ampere (A); mole (mol); and candela (cd).

The earlier literature references and some of the older text books, naturally use the older units. Even now many practicing scientists have not adopted the SI unit as their working unit. It is therefore necessary to know of the older units and be able to interconvert with SI units.

In this series of texts SI units are used as standard practice. However in areas of activity where their use has not become general practice, eg biologically based laboratories, the earlier defined units are used. This is explained in the study guide to each unit.

Table 1 shows some symbols and abbreviations commonly used in analytical chemistry; Table 2 shows some of the alternative methods for expressing the values of physical quantities and the relationship to the value in SI units.

More details and definition of other units may be found in the *Manual of Symbols and Terminology for Physicochemical Quantities and Units*, Whiffen, 1979, Pergamon Press.

Table 1 *Symbols and Abbreviations Commonly used in Analytical Chemistry*

Å	Angstrom
$A_r(X)$	relative atomic mass of X
A	ampere
E or U	energy
G	Gibbs free energy (function)
H	enthalpy
J	joule
K	kelvin (273.15 + t °C)
K	equilibrium constant (with subscripts p, c, therm etc.)
K_a,K_b	acid and base ionisation constants
$M_r(X)$	relative molecular mass of X
N	newton (SI unit of force)
P	total pressure
s	standard deviation
T	temperature/K
V	volume
V	volt (J A^{-1} s^{-1})
a, a(A)	activity, activity of A
c	concentration/ mol dm^{-3}
e	electron
g	gramme
i	current
s	second
t	temperature / °C
bp	boiling point
fp	freezing point
mp	melting point
$\approx$	approximately equal to
$<$	less than
$>$	greater than
e, exp(x)	exponential of x
ln x	natural logarithm of x; ln x = 2.303 log x
log x	common logarithm of x to base 10

Table 2 *Alternative Methods of Expressing Various Physical Quantities*

1. **Mass (SI unit : kg)**

$$g = 10^{-3}\ kg$$
$$mg = 10^{-3}\ g = 10^{-6}\ kg$$
$$\mu g = 10^{-6}\ g = 10^{-9}\ kg$$

2. **Length (SI unit : m)**

$$cm = 10^{-2}\ m$$
$$Å = 10^{-10}\ m$$
$$nm = 10^{-9}\ m = 10Å$$
$$pm = 10^{-12}\ m = 10^{-2}\ Å$$

3. **Volume (SI unit : m^3)**

$$l = dm^3 = 10^{-3}\ m^3$$
$$ml = cm^3 = 10^{-6}\ m^3$$
$$\mu l = 10^{-3}\ cm^3$$

4. **Concentration (SI units : mol m^{-3})**

$$M = mol\ l^{-1} = mol\ dm^{-3} = 10^3\ mol\ m^{-3}$$
$$mg\ l^{-1} = \mu g\ cm^{-3} = ppm = 10^{-3}\ g\ dm^{-3}$$
$$\mu g\ g^{-1} = ppm = 10^{-6}\ g\ g^{-1}$$
$$ng\ cm^{-3} = 10^{-6}\ g\ dm^{-3}$$
$$ng\ dm^{-3} = pg\ cm^{-3}$$
$$pg\ g^{-1} = ppb = 10^{-12}\ g\ g^{-1}$$
$$mg\% = 10^{-2}\ g\ dm^{-3}$$
$$\mu g\% = 10^{-5}\ g\ dm^{-3}$$

5. **Pressure (SI unit : N m^{-2} = kg m^{-1} s^{-2})**

$$Pa = Nm^{-2}$$
$$atmos = 101\,325\ N\ m^{-2}$$
$$bar = 10^5\ N\ m^{-2}$$
$$torr = mmHg = 133.322\ N\ m^{-2}$$

6. **Energy (SI unit : J = kg m^2 s^{-2})**

$$cal = 4.184\ J$$
$$erg = 10^{-7}\ J$$
$$eV = 1.602 \times 10^{-19}\ J$$

Table 3 *Prefixes for SI Units*

Fraction	Prefix	Symbol
10^{-1}	deci	d
10^{-2}	centi	c
10^{-3}	milli	m
10^{-6}	micro	μ
10^{-9}	nano	n
10^{-12}	pico	p
10^{-15}	femto	f
10^{-18}	atto	a

Multiple	Prefix	Symbol
10	deka	da
10^{2}	hecto	h
10^{3}	kilo	k
10^{6}	mega	M
10^{9}	giga	G
10^{12}	tera	T
10^{15}	peta	P
10^{18}	exa	E

Table 4 *Recommended Values of Physical Constants*

Physical constant	Symbol	Value
acceleration due to gravity	g	9.81 m s^{-2}
Avogadro constant	N_A	$6.022\ 05 \times 10^{23} \text{ mol}^{-1}$
Boltzmann constant	k	$1.380\ 66 \times 10^{-23} \text{ J K}^{-1}$
charge to mass ratio	e/m	$1.758\ 796 \times 10^{11} \text{ C kg}^{-1}$
electronic charge	e	$1.602\ 19 \times 10^{-19} \text{ C}$
Faraday constant	F	$9.648\ 46 \times 10^{4} \text{ C mol}^{-1}$
gas constant	R	$8.314 \text{ J K}^{-1} \text{ mol}^{-1}$
'ice-point' temperature	T_{ice}	273.150 K exactly
molar volume of ideal gas (stp)	V_m	$2.241\ 38 \times 10^{-2} \text{ m}^3 \text{ mol}^{-1}$
permittivity of a vacuum	ϵ_o	$8.854\ 188 \times 10^{-12} \text{ kg}^{-1} \text{ m}^{-3} \text{ s}^4 \text{ A}^2 \text{ (F m}^{-1})$
Planck constant	h	$6.626\ 2 \times 10^{-34} \text{ J s}$
standard atmosphere pressure	p	$101\ 325 \text{ N m}^{-2}$ exactly
atomic mass unit	m_u	$1.660\ 566 \times 10^{-27} \text{ kg}$
speed of light in a vacuum	c	$2.997\ 925 \times 10^{8} \text{ m s}^{-1}$

BARTON COLLEGE LIBRARY
543.0896 W669g nonf
Willett, John/Gas chromatography
3 6500 00052 9955
DISCARDED BARTON LIBRARY
Barton College Library
Wilson, N.C. 27893